# WORLDWATCH REPORT 176

# Farming Fish for the Future

BRIAN HALWEIL

LISA MASTNY, *EDITOR*

WORLDWATCH INSTITUTE, WASHINGTON, DC

© Worldwatch Institute, 2008
ISBN 978-1-878071-85-9
Library of Congress Control Number: ———X

Printed on paper that is 50 percent recycled, 30 percent
post-consumer waste, process chlorine free.

*On the cover:* A web of salmon, mussels, and kelp.

Illustration © 2008 by Joan A. Wolbier

## Table of Contents

### Figures, Tables, and Sidebars

## Acknowledgments

I am grateful to the many people who shared their knowledge and criticism for this report, including Malcolm Beveridge, Margaret Bowman, Jim Carlberg, Thierry Chopin, Barry Costa-Pierce, Stephen Cross, Steven Damato, Sena De Silva, Chad Dobson, Phaedra Doukakis, Rebecca Goldburg, Chuck Hesse, Teresa Ish, Daniel Lee, Patricia Majluf, Chris Mann, Roz Naylor, Ellen Pikitch, Shawn Robinson, Henning Roed, Astrid Scholz, Neil Sims, Don Staniford, Albert Tacon, John Volpe, and Helene York.

At Worldwatch, several colleagues provided thoughtful feedback, including Danielle Nierenberg and Robert Engelman. Intern Hannah Doherty helped with last-minute fact checking and contributed a sidebar on the healthfulness of farmed fish. Intern Amanda Chiu helped to update all of the data series, and Staff Writer Ben Block contributed a sidebar on a Baltimore-based project to raise fish in tanks.

Senior Editor Lisa Mastny trimmed down a monstrous early draft, allowing the most relevant points and examples to shine and asking questions that helped turn the unclear into the insightful. Art Director Lyle Rosbotham transformed dry data into clear graphics and selected supporting images, including cover art by his wife Joan Wolbier. And the Communications team of Darcey Rakestraw and Julia Tier, and Marketing Director Patricia Shyne, helped tease out the most important messages for Worldwatch's varied audience.

The W.K. Kellogg Foundation provided a grant that supported this work.

## About the Author

**Brian Halweil**, a senior researcher, joined Worldwatch in 1997 as the John Gardner Public Service Fellow from Stanford University. Brian writes on the social and ecological impacts of how we grow food, focusing recently on organic farming, hunger, seafood, and rural communities. He described the evolving local food movement in his 2004 book *Eat Here: Reclaiming Homegrown Pleasures in a Global Supermarket*, and highlighted the important role of consumer demand for "sustainable seafood" in his 2006 report *Catch of the Day: Choosing Seafood for Healthier Oceans*.

Brian's work has been featured in the international press, and he has testified before the U.S. Senate Committee on Foreign Relations on the role of biotechnology in combating poverty and hunger in the developing world. He has traveled extensively in Mexico, Central America and the Caribbean, and East Africa learning indigenous farming techniques and promoting sustainable food production. Before coming to Worldwatch, Brian worked with California farmers interested in reducing their pesticide use and set up a two-acre student-run organic farm on the Stanford campus. He writes from Sag Harbor, New York, where he and his family tend a home garden and orchard and rake for clams when the tide allows.

# Summary

From Asia to North America, people are eating more seafood, either because it's the most affordable form of protein (as in many poorer nations) or because it's the latest health food trend (as in many wealthy nations). But as the demand for fish rises, populations of both marine and freshwater species are being over-exploited, resulting in stagnant or declining catches from many wild fisheries.

As a result, seafood is shifting from being the last wild ingredient in our diet to being a highly farmed commodity. Farmed seafood, or aquaculture, now provides 42 percent of the world's seafood supply and is on target to exceed half in the next decade. Fish farms are taking up more space on land and at sea, as farmers expand into new streams, bays, and oceans. Fish farming itself has morphed from a small-scale, artisanal pursuit into a large-scale science, with innovations in feed technology, cage design, and fish breeding.

Farmed seafood has certain advantages over wild fish in meeting modern demand. For a global marketplace that demands increasingly predictable products—uniform-sized fillets available year-round, free of the vagaries of weather or open-ocean fishing—fish farming delivers this predictability. Farms are also becoming more productive, raising fish at a lower cost and expanding the potential market.

Yet even as we depend more on farmed fish, several crises loom that may jeopardize future expansion of this industry. These include a growing scarcity of fish feed and rising concern about the social and ecological fallout from industrial aquaculture. Poorly run fish farms can generate coastal pollution in the form of excess feed and manure, and escaped fish and disease originating on farms can devastate wild fisheries. From salmon farms in Chile to tilapia farms in China, a narrowing base of genetic diversity means that farms will be increasingly susceptible to disease and other stresses, a well-known pattern in agriculture that may play out in aquaculture.

But not all fish farming is created equal. Still today, most aquaculture is focused on sea-weeds, shellfish, and other species that are low on the food chain, such as carp and tilapia. For much of the world, particularly the developing world, fish farming isn't so much about profit as about having a steady supply of seafood to eat. Most fish farms are small in scale, rely on few inputs, and may be closely integrated with crop or livestock production. From the Philippines to Bangladesh to the southern United States, small-scale fish farmers often have higher and more stable incomes than nearby crop farmers.

Yet the greatest growth in fish farming today is occurring at the other end of the spectrum: large farms raising high-value, predatory fish such as salmon, striped bass, tuna, and shrimp. Raising these species is an exercise in "reducing" fish to produce fish—that is, in turning certain fish, usually smaller species such as anchovy, herring, capelin, and whiting, into feed for other, typically larger, species. Increasingly, we are fishing *down* the ocean chain so we can move *up* the fish-farming chain.

Despite ongoing improvements in feed ingredients and technologies, the rapid growth in fish farming in recent decades has effectively outweighed any gains in feeding efficiency. According to most estimates, modern fish farming is now a net drain on the world's seafood supply. The global appetite for farmed fish is putting unsustainable strain on the world's food resources.

As farmers raise more predatory species, a focus on well-designed fish farms will make a critical difference. To avert the looming feed crisis and to take pressure off perfectly edible wild fish, farmers could wean themselves off fish-based feed. And some innovative fish farmers are beginning to redesign their farms to function more like healthy aquatic ecosystems. Farms with high levels of integration can greatly reduce water pollution and disease levels. They can be a cost-effective way to recycle, clean, and store water supplies. They can even help rebuild wetlands and restock wild fisheries.

Properly guided, the explosive growth in fish farming may in fact be the most hopeful trend in the world food system. Compared to raising cows, pigs, or even chicken, aquaculture is remarkably efficient in its use of feed and water. And farmed fish are still generally lower on the food chain and less resource-intensive than the big predatory fish we catch in the seas. Rather than contributing to environmental degradation, fish farming can be a critical way to add to the global diet.

Yet there is no guarantee that aquaculture will move wholesale in a "greener" direction. Supportive government policies and a shift in consumer tastes will be essential to push farmers toward raising more-efficient species, such as carp, catfish, and shellfish. The seafood and aquaculture industries must also play a significant role. So far, producers and conservation groups have only begun discussing standards for farmed fish, despite a proliferation of eco-labels for wild seafood and other agricultural products. Without such standards, even concerned seafood eaters won't be able to push the world's fish farms in the right direction.

# A Different Sort of Salmon

It's hard to imagine that, standing on the shores of the quaint coastal community of Back Bay in southwestern New Brunswick, you can see the future of the world's seafood.

Sprawled out before you are the fish cages of Cooke Aquaculture, the largest salmon producer in eastern Canada. Most of the operation is underwater, with a handful of bobbing blue buoys the only evidence of what's going on below. These waters are part of the Bay of Fundy, renowned for the greatest tidal swings on Earth—as much as 10 meters of vertical change. The swings make this site particularly good for aquaculture, which depends in part on good water flow.[1]*

But don't just think of this place as a productive and profitable farming operation, which it is. Think of it as an experiment in ecosystem creation.

Most other salmon farms, including many of Cooke's operations elsewhere in Canada, have raised the ire of local residents and waterfront activists because of the high levels of feed and manure that leak into surrounding waters. But this farm has dramatically reduced its pollution. That's because it doesn't just raise salmon. Mimicking some of the functions of a coastal ecosystem, it raises three different species that end up complementing each other in important ways.

At the center of the operation is a cluster of 15 large salmon cages, each running 70 meters on end. The metal cages are surrounded by four additional units that have been retrofitted to hold socks of blue mussels, which drape down into the water column. The shellfish,

positioned some 20 meters "downstream" from the salmon, function as a filter for any excess waste flowing from the fish cages. Still farther out, beyond the mussels, is a flotilla of large rafts from which dangle long ribbons of kelp. The seaweed thrives on the dissolved nitrogen, phosphorus, and other nutrients that diffuse into the water from the salmon operation.

"We have to redesign [fish] farms so that they will support the cultivation of several species," says Thierry Chopin, a biology professor at the University of New Brunswick who has conducted research at Cooke's sites.[2] Moving to Canada from France in 1989, Chopin was concerned about the "nutrient enrichment" from local fish farms—in particular the pollution of surrounding waters. He wondered if seaweeds could help fix the leaks. Around that time, he met Shawn Robinson, a scientist with the fisheries department in St. Andrews, who was wondering the same thing about shellfish. "We both realized that we worked on extractive species," said Chopin. "And we said, 'How can we combine them with fed species?'"[3] In Chopin and Robinson's world, fish waste isn't simply a source of pollution; it's a wasted source of energy for seaweed, shellfish, and ultimately humans.

This innovative approach to raising fish comes at a critical time. From Asia to North America, people are eating more seafood, either because it's the most affordable form of protein (as in many poorer nations) or because it's the latest health food trend (as in many wealthy nations). On average, each person on the planet is eating four times as much seafood as was consumed in 1950.[4] As the demand for fish rises, populations of both marine and

---

*Endnotes are grouped by section and begin on page 37.

freshwater species are being overexploited, resulting in stagnant or declining catches from many wild fisheries.[5]

Seafood is shifting from being the last wild ingredient in our diet to being a highly farmed commodity. And fish farming itself has morphed from a small-scale, artisanal pursuit into a large-scale science. With advances in breeding, feed formulations, and pen design, farms are becoming more efficient and more productive, allowing for an increasingly predictable product that is available year-round, free from the vagaries of weather or open-ocean fishing.[6] In contrast to a swordfish caught at sea, which may need to be stored on ice for several days before it can be processed, a farmed salmon or tilapia can be harvested at will and taken immediately to a processing plant to be filleted, breaded, seasoned, frozen, and packaged for an infinite range of consumer desires.[7]

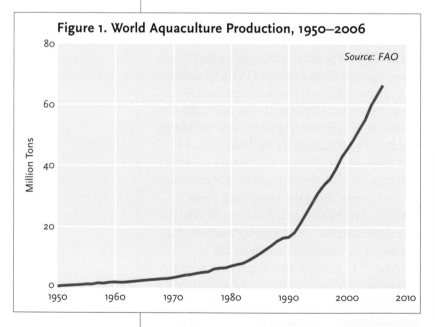

**Figure 1. World Aquaculture Production, 1950–2006**

Source: FAO

These advantages are part of the reason that farmed seafood, or aquaculture, constitutes a greater share of the world's seafood supply than ever before. It now provides 42 percent of the total and is on target to exceed half in the next decade.[8] In 2006, fish farmers raised an estimated 66.7 million tons of seafood worth more than $80 billion—nearly double the volume of a decade earlier.[9]* (See Figure 1.) Put in a dietary context, the average per capita con-

sumption of farmed seafood has increased nearly tenfold since 1970, to more than 10 kilograms in 2006.[10] In contrast, per capita meat consumption grew just 60 percent, to 43 kilograms.[11]

Fish farms are taking up more space on land and at sea, as farmers expand into new streams, bays, and oceans and install more ponds, fish pens, shellfish cages, and seaweed rafts.[12] Experts predict that farmed seafood will grow an additional 70 percent by 2030.[13] In the United States, currently a minor contributor to global aquaculture, the Department of Commerce has adopted a policy to increase the value of domestic production fivefold by 2025.[14]

Yet even as we depend more on farmed fish, several crises loom that may jeopardize future expansion of this industry. These include a growing scarcity of fish feed and rising concern about the social and ecological fallout from industrial aquaculture. Poorly run fish farms can generate coastal pollution in the form of excess feed and manure, and escaped fish and disease originating on farms can devastate wild fisheries.

Marine scientists, culinary experts, and others worry about about the implications of "taming" the last wild ingredient in the global diet. Compared to the relatively slow process of domesticating land plants and livestock over thousands of years, fish farmers are domesticating mollusks, fish, jellyfish, and sea plants roughly 100 times more rapidly.[15] From salmon operations in Chile to tilapia farms in China, a narrowing base of genetic diversity means that farms will be increasingly susceptible to disease and other stresses, a well-known pattern in agriculture that may play out in aquaculture.[16]

But not all fish farming is created equal. So far, most of the world's aquaculture is focused on seaweeds, shellfish, and other species that are low on the food chain, such as carp and tilapia.[17] (See Figure 2.) Yet as farmers raise more predatory (and resource-intensive) species, such as salmon and tuna, a focus on well-

---

* Units of measure throughout this report are metric unless common usage dictates otherwise.

designed fish farms will make a critical difference. Farms with high levels of integration, such as Cooke's Back Bay project, can greatly reduce water pollution and disease levels. They can be a cost-effective way to recycle, clean, and store water supplies. They can even help rebuild wetlands and restock wild fisheries.

mussels grown on their own.[20] And because salmon stocked at high density tend to develop more health problems in murky, nutrient-clogged water, the cleaner, mixed system means lower veterinary costs. There is also evidence from Back Bay, as well as from farms in Norway and Maine, that mussels may help reduce the

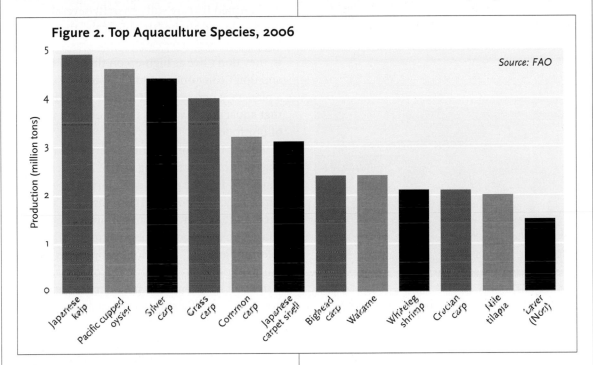

**Figure 2. Top Aquaculture Species, 2006**

Source: FAO

Such systems—dubbed IMTA, or integrated multitrophic aquaculture, but more easily called ecological aquaculture—remain in their infancy, though they are spreading. And operations like Cooke's are leading the way. Because the farm's mussels and kelp don't capture all of the waste generated by the salmon, Chopin and Robinson have considered adding sea urchins or sea cucumbers to remove the bigger particles.[18] In their view, a more efficient, integrated system could include as many species as possible, serving as many functions as possible. Chopin even envisions using the waste from an inshore salmon hatchery, where the fish are raised from eggs before being placed in the cages, as fertilizer in an adjoining aquaponic vegetable operation—an approach being used successfully on a small scale elsewhere.[19]

There are other advantages to the integrated system. The mussels grow faster and supply nearly 40 percent more meat per shell than

virulence of infectious salmon anemia (ISA), a devastating disease that requires farmers to use costly antibiotics or adopt even more expensive quarantine procedures that can shut down entire farms.[21] (Despite these benefits, however, some local groups continue to oppose expanded aquaculture in the Bay of Fundy because of concerns about other forms of "pollution," including sea lice infestation, fish escapes, and the transfer of disease to wild stocks.[22])

Because the mussels and seaweed don't have to be fed, they are relatively low maintenance for most of the year. Their upkeep is less time-sensitive than raising fish and can be done during the downtime in salmon work. Robinson sees integrated systems like Back Bay's as "a bridging mechanism between traditional aquaculture and environmental NGOs"—a form of fish farming that any conservationist should love.[23] The resulting seafood might even carry ecolabels, allowing producers to sell it at a premium.

But Cooke Aquaculture has an even larger vision. "The environmental aspect is…very important as a hedge against future regulation," Chopin explains.[24] Eventually, when fish farms are required to take responsibility for

Ecological aquaculture in action: salmon cages on the left, a mussel raft in the right foreground, and kelp rafts in the right background. This is one of Cooke Aquaculture's integrated salmon farming operations.

Courtesy Theirry Chopin

their waste, as is likely in the coming decades, Cooke will be ahead of the pack. The company, which has around 100 salmon farming sites in eastern Canada, is already using the integrated technique at six sites and plans to add several new ones each year.[25] "They have done their calculations and they believe in the concept from environmental, social, and economic advantages," says Chopin. "These guys don't do it just for fun."[26]

Properly guided, the explosive growth in fish farming may in fact be the most hopeful trend in the world food system. Compared to raising cows, pigs, or even chicken, aquaculture is remarkably efficient in its use of feed and water, in part because fish burn less energy resisting gravity than most land animals do.[27] And farmed fish are still generally several notches lower on the "trophic scale" than the big predatory fish we catch in the seas—that is, they don't live as high up on the food chain and don't consume as many resources. So from a planetary perspective, it may be a good thing that aquaculture accounts for more and more of our food.

Rather than contributing to environmental degradation, fish farming can be a critical way to add to the global diet, particularly as a hedge against potential crop losses or shortages in the supply of meat. Yet there is no guarantee that aquaculture will move wholesale in a "greener" direction. Supportive government policies and a shift in consumer tastes will be essential to push farmers toward raising more-efficient species, such as carp, catfish, and shellfish. The seafood and aquaculture industries must also play a significant role. So far, producers and conservation groups have only begun discussing standards for farmed fish, despite a proliferation of eco-labels for wild seafood and other agricultural products. Without such standards, even concerned seafood eaters won't be able to push the world's fish farms in the right direction.

# From Ornamental Ponds to Industrial Aquaculture

According to archaeologists, fish farming first took root several thousand years ago. With the growth in human populations and a decline in more-accessible fish supplies, it became harder for people to rely entirely on wild-caught fish, shellfish, and seaweeds.[1] To increase the supply of wild fish, communities would take steps such as transplanting already-fertilized eggs, trapping fish in shallow or dammed areas, and generally creating favorable fish habitats by excluding predators and feeding wild species.[2]

As with land-based agriculture, aquaculture emerged chiefly in settled societies. Early examples include tilapia raising in Egypt some 4,000 years ago, the *chinampas* system in Aztec Mexico (where fish thrived in irrigation canals next to crop beds) around 1200 A.D., and the treatise on carp farming written by Chinese scholar Fan Li some 2,500 years ago.[3] Clay models found in graves dating from China's Han Dynasty depict rice fields dotted with 18 varieties of aquatic plants and animals that are still used in the country today, from the now-ubiquitous carp to lotus flowers and soft-shelled turtles.[4]

Fish farming made sense in China historically because of the large human population, an agricultural landscape that was too crowded for extensive livestock raising, and the ubiquity of rice cultivation in low-lying, often flooded deltas, which created a natural setting for fish.[5] From roughly 200 B.C. to 200 A.D., the practice graduated from rice paddies and ponds to lakes, a shift that would require the use of cages, pens, or other enclosures.[6] Around 600 A.D., Chinese farmers began raising multiple species of carp together to make better use of

This ornamental koi pond is outside a Buddhist temple on the island of Oahu, Hawaii.

LeoSynapse

feeds and to boost total production.[7]

The first form of mixing crops, livestock, and fish was China's fishpond-dyke-mulberry system, in which fruit trees were planted on the dykes separating fish ponds.[8] The tree leaves and fruit would partly feed the fish, while the ponds irrigated the tree roots. Farmers in 14th century Europe evolved a similarly sophisticated system of rotating three seasons of agricultural crops with three fish crops, based on the understanding that the pond sediment from fish manure could serve as an excellent fertilizer for livestock pastures.[9]

At times, early aquaculture served purposes other than providing food, such as using the ponds to create a year-round water supply or waste disposal site, or cultivating the fish for entertainment and aesthetic purposes.[10] In Medieval Europe, it wasn't uncommon for castle latrines to drain directly into fish ponds, where the nutrients in the waste would feed the plant life or fish. The need for a steady supply of fish for religious purposes in Buddhist Japan, Ancient Egypt, Christian Europe, and elsewhere also pushed the development of aquaculture.[11]

In Asia, fish farming has been a natural addition to rice farming for thousands of years. Vegetable scraps and crop residues are fed to the fish, and the fish produce waste that is used to fertilize the rice fields.[12] The system enables farmers to save money on pesticides and herbicides, since the fish help control pests by consuming the larvae, weeds, and algae that carry disease and compete with the rice for nutrients.[13] (Fish farming also helps control

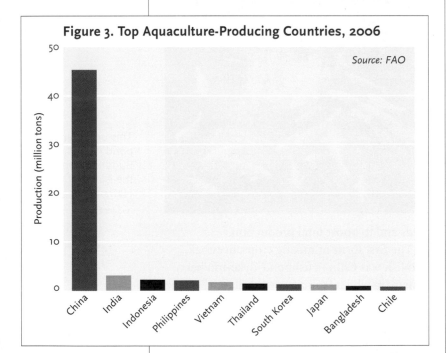

**Figure 3. Top Aquaculture-Producing Countries, 2006**

Source: FAO

Production (million tons)

China, India, Indonesia, Philippines, Vietnam, Thailand, South Korea, Japan, Bangladesh, Chile

malaria, since fish eat mosquito larvae.[14])

But what distinguishes this traditional fish farming, and all fish farming until the 20th century, is the raising of herbivorous fish, fed on vegetable scraps rather than on other fish. As a result, these early techniques increased the overall supply of seafood, with minimal impact on the environment and on wider fish populations. There were rare exceptions, of course: in Ancient Rome, the occasional tuna or other carnivorous fish was trapped in coastal cages

for the benefit of the nobility, but only great wealth allowed this.[15]

The Chinese were the first to systematically breed fish and control fish spawning, but fish farming didn't really take off in the country until the founding of the People's Republic in 1949. As a lesser-known outcome of Mao's push for self-sufficiency, the government constructed some 82,000 artificial water bodies for hydropower, flood control, and irrigation. This added more than 2 million hectares of inland water surface in China and led to a 50-fold expansion in aquaculture nationwide.[16] Today, China produces nearly 70 percent of global farmed seafood.[17] (See Figure 3.) More than three-quarters of the country's food fish comes from aquaculture, compared with just 20 percent for the rest of the world.[18]

China's decisions over the next few years will dramatically shape the future of fish farming.[19] In addition to being the world's largest producer of farmed fish, the country is the largest consumer of fish feed as well as the largest importer of fishmeal and fish oil, two key sources of protein and fats for animal feeds. China is also the largest producer of predatory fish, including black carp, river eels, and marine shrimp, raising more than 1.5 million tons annually, or about 30 percent of global production.[20] These carnivorous species so far comprise only a small share of the country's total aquaculture output of roughly 45 million tons, but a strong export market and domestic appetite mean that production is growing rapidly.[21]

Many Chinese fish farmers are now supplementing the traditional plant-based diet fed to carp with fishmeal, or feed made from other fish. This could generate tremendous new demand for the feed, given that 16 million tons of carps are raised annually worldwide.[22] Observers note that across China, "traditional aquaculture practices are being abandoned in favor of intensification." The use of night soil (human manure) to feed ponds, for example, is being pushed out both by social stigma and by the adoption of sewage containment systems that mix industrial and human waste.[23]

Not just in China, but around the world,

## Sidebar 1. Factory Farming at Sea

Much like the move to concentrated factory farms to produce meat, fish farmers are increasingly raising large numbers of fish in close proximity. This "industrial aquaculture" generates significant amounts of waste that can pollute nearby waters. A fish farm with 200,000 salmon releases nutrients and fecal matter roughly equivalent to the raw sewage from 20,000 to 60,000 people (though fish waste is not as dangerous in terms of pathogens as human or livestock waste). Scotland's salmon aquaculture industry is estimated to produce the same amount of nitrogen waste as the untreated sewage of 3.2 million people, or just over half the country's total population. In sites where there is little flushing by tides and currents, the waste from net pen operations can create a dead zone on the ocean floor that extends from 30 to 150 meters in diameter.

The cramped conditions in these facilities are hard on the fish and also encourage the spread of disease, so fish farmers sometimes rely on antibiotics, de-licing compounds, and other chemicals, most of which end up in the water. Fish farmers spend nearly $1 billion each year on veterinary products. And losses to disease are often in the hundreds of millions of dollars. In recent years, shrimp farmers in China have lost $120 million to bacterial fish diseases and $420 million to shrimp diseases.

Many of these diseases, such as sea lice, are also present in wild fish populations, but natural migration patterns typically ensure that an entire school isn't decimated. In the case of wild salmon, adult fish and juveniles spend part of their lives apart, so the juveniles have a chance to build up immunity to the lice before encountering the lice-harboring adults. The development of salmon farms in coastal waters, however, means that wild juvenile salmon often migrate directly through cages of infected farmed salmon.

This susceptibility of farmed fish to disease is a growing concern as farmers raise a wider diversity of species. "There are a dozen diseases or organisms that could be just as devastating as sea lice," explains John Volpe, an aquaculture specialist at the University of Victoria in British Columbia. Volpe says the existing experience with lice and salmon farms should give the Canadian province pause about launching a new sablefish farming industry, as long as sablefish is also an important wild fishery.

As aquaculture becomes increasingly globalized, with fish stocks, feeds, and other inputs moving around the world, the potential for transporting disease also grows. In 1995, a herpes virus erupted near tuna ranches off southern Australia and spread through regional waters at a rate of 30 kilometers a day. The epidemic left 75 percent of the pilchard population dead and triggered a mass starvation of seabirds such as gannets and penguins. It was one of the largest mass mortalities ever reported for marine species. Although the virus's origins remain unknown, some analysts blame the tuna industry's growing dependence on imported feed, which now accounts for nearly 30 percent of all fish fed to Australian tuna.

An outbreak on a similar scale has not occurred since, although the movement of fish and feed across borders continues. Still, the crisis did get the industry thinking about biosecurity and the wisdom of moving feed from one ocean to another. Halting the spread of such disease will only become more difficult as the international trade in fishmeal—often from multiple sources, countries, and species, and combined in a single mixture—continues to grow.

*Source: See Endnote 24 for this section.*

fish farming is rapidly following the path of livestock production, with large operations raising massive numbers of genetically uniform animals.[24] (See Sidebar 1.) While much of the farming of carp, tilapia, shellfish, and seaweed across Asia shares principles with ancient low-input techniques, modern farms that raise predatory species such as salmon and shrimp draw most directly on industrial principles. They are removed from ecological systems, dependent on external inputs, and generate large amounts of waste. The genetic uniformity of farmed shrimp and salmon is already inviting large disease outbreaks and dependence on antibiotics—not too different from the dependence of livestock production on similar medications.

In many ways, aquaculture is following the same trajectory as land-based agriculture, but over a dramatically shorter timespan. On land, farming has been the main source of human food acquisition for some 12,000 years, ever since the shift from hunting and gathering during the Agricultural Revolution. In the water, however, fishing for seafood has always dominated, in part because wild fish were abundant and because the cultivation of water-dwelling animals proved less intuitive for humans.[25] Even in China, with its long history of fish farming, aquaculture was generally only a complement to catching wild fish—a side business for rice farmers or coastal communities.

It is only in the last few decades, as production from fish farms has grown at double-digit rates, that farmed seafood has begun to compete with wild seafood.[26] Yet this tremendous growth has come with almost no guidance. It's as if we've hardly begun to consider fish farming seriously, even though it will soon account for half of our seafood.

Salmon is perhaps the best example of fish farming's evolution. Europeans first started raising salmon in hatcheries in the late 1700s, releasing the young fish to rivers in the hopes of enhancing wild runs that had been depleted.[27] But it wasn't until the 1970s that farmers began to raise salmon entirely in captivity. Norway quickly emerged as the world leader, with production expanding rapidly in the 1980s. The practice then spread to Scotland, Japan, Chile, Canada, the United States, Ireland, New Zealand, Australia, and the Faroe Islands.[28] As recently as 1980, farmed salmon accounted for only about 1 percent of global salmon output; but by the early 1990s, nearly twice as much farmed salmon was harvested as wild salmon.[29]

### Table 1. Changes in Price and Production, Selected Aquaculture Species

| Product | Period | Price Decline | Production Increase |
|---|---|---|---|
| | | (percent) | (percent) |
| Atlantic salmon | 1986/87–2004 | 20–40 | 3,108 |
| Pacific white-leg shrimp | Recent | 62 | 854 (2000–04) |
| Japanese eel | 1988–2004 | 71 | 159 |
| Common carp | 1984–2004 | 40 | 397 |
| Tilapia | 1992–2004 | 20 | 164 |

*Source: See Endnote 34 for this section.*

Many of the earliest Norwegian salmon farmers were sailors who had witnessed fish farming in Japan and in other parts of Asia and Europe, according to Henning Roed, a marine biologist in Oslo who has worked with both

the government and fish farmers.[30] "They made a lot of mistakes," Roed says, including copying rainbow trout operations they had seen in Germany and Denmark and confining the salmon in dug-out inland areas. The farmers received little institutional support, and their fish barely stayed alive.[31]

Then people from three related fields got involved, Roed explains. First, fishermen and hatchery experts suggested that salmon may do better in pens and cages in the water. Second, people with a farming background realized that salmon raised in confinement would be susceptible to diseases and would need to be treated, just like livestock. Finally, once the basic production challenges were overcome, entrepreneurs with business know-how, marketing prowess, and connections to government subsidies and university research got involved. Soon, salmon began muscling out beef, pork, and other meats at the center of the Norwegian diet. "Now it's like the chicken of the ocean," says Roed.[32]

As salmon shifted from being a seasonal, wild product of high value to being a farmed commodity, this brought a new set of economic, social, and ecological consequences. One immediate effect was that the surge in farm production pushed down the prices of both wild and farmed product.[33] Between 1988 and 2002, the price of farmed Atlantic salmon dropped 61 percent, and prices for Pacific salmon species that compete most highly with Atlantic species fell 59–64 percent. Similar declines have been seen for other major farmed species, from oysters and white-leg shrimp to carp and tilapia.[34] (See Table 1.)

This doesn't just mean a blow to the livelihood of wild fishers. It also means that fish farmers have to compete on volume to make up for falling profit margins. In the salmon industry and other fish-farming sectors, successive waves of consolidation have led farms and aquaculture companies to cannibalize their smaller brethren, often to be cannibalized at a later date by someone even larger.[35]

Ownership in industrial aquaculture has become highly concentrated. In 2001, just 30 or so companies controlled two-thirds of the world's farmed salmon and trout production.[36] The larger firms—multibillion-dollar companies like Marine Harvest (which recently merged with its major competitor, Panfish, which had itself gobbled competitor Fjord Seafood) and Cermaq—operate in multiple countries and often control the entire production process, from the feed, hatchery, and grow-out to processing and distribution.[37] For salmon, more than two-thirds of global aquafeed production is overseen by just two companies, Skretting (Nutreco) and Ewos (Cermaq).[38] As in agribusiness more generally, such concentration does not bode well for meeting the needs of hungry people.

As with other agricultural commodities, the profits and jobs from fish farming do not always stay where the fish are farmed.[39] In the case of British Columbia's salmon industry, as the price of the fish fell, so did wages on salmon farms and in processing plants, and jobs in commercial fisheries also dried up.[40] In the early 1980s, the province was home to 75 salmon-farming companies, mostly smaller outfits. Today, just two companies control more than 80 percent of production.[41] And export earnings have not increased significantly with the introduction of salmon farming, in part because of a global glut from overproduction.[42]

"The most damaging aspect of the industrial salmon farming experience is the establishment of a consumer culture expecting that salmon should be available fresh year-round for the same price as chicken," notes University of Victoria aquaculture specialist John Volpe, who calls the fish "battery chickens of the sea."[43] Volpe compares "cheap" salmon to "cheap" Amazonian beef, arguing that the main reason these foods are so inexpensive is because the costs are transferred to the environment and to society at large. These costs include the heavy usage of clean, oxygenated water; the "removal" of organic and other wastes by natural currents; the assimilation of escaped fish into wild populations; and high levels of sea lice in the ecosystem.[44]

Salmon isn't the only fish that has been transformed by fish breeding, know-how adapted from livestock farming (especially in the development of fish feed), and veterinary techniques for helping fish thrive.[45] Farmed production of many other species is also beginning to approach or exceed wild production. Wild landings of Atlantic halibut, a large, bottom-dwelling flat fish, have declined precipitously; farmed production now equals roughly 10 percent of the wild catch worldwide and 25 percent in Norway.[46] With some species, farmed production is being shifted to open-ocean environments to further increase output and to operate "out of sight" of coastal residents.[47] (See Sidebar 2.)

### Sidebar 2. Open-Ocean Aquaculture

Fish farms are moving deeper into the open ocean, since coastal communities don't want to see the farms or deal with the pollution that can result. In the open ocean, submersible cages are anchored to the seafloor but can be moved within the water column, enabling them to avoid rougher surface waters and minimizing interference with ships and other vessels. The cages, which are large enough to hold hundreds of thousands of fish, are tethered to buoys that control feeding; robots may be used to inspect, clean, or monitor the cages.

Future ocean-based operations could become even more sophisticated. As researchers Roz Naylor and Marshall Burke observe, "The next generation technology...includes a gigantic cage that will travel hundreds of miles offshore and roam the seas instead of remaining fixed to a buoy. Juvenile tuna placed in roaming cages in Mexico could conceivably arrive in Japan ready for market sales several months later."

But realizing that their customers—and ocean activists—are increasingly savvy about how fish are farmed, these offshore farms are constantly and aggressively changing their practices. Kona Blue, a cobia farm located roughly a kilometer off the coast of Hawaii, has openly responded to its critics. In fact, questions from the Environmental Defense Fund and other groups actually inspired Kona's evolution toward better monitoring of the seawaters surrounding the operation. The farm also replaced some of its fishmeal with vegetable products, and specifically farms a species native to the area. "It's not like there's an endpoint to this discussion," says marine biologist Neil Sims, the farm's co-founder. "We are always looking to improve."

Critics have lingering concerns about open-ocean operations. First, they argue that waste plumes coming from aquaculture aren't harmlessly dispersed in the water. Second, these farms are not immune to the forces of nature, and massive storms could mean large escapes of fish, including some species that don't currently inhabit certain parts of the oceans. Finally, this production is expensive. "It's not feeding the world. It's about trying to make money," says Chris Mann, senior officer and director of the Campaign for Healthy Oceans at the Pew Environment Group.

*Source: See Endnote 47 for this section.*

A salmon farm floating in Loch Diabaig, on the western coast of Scotland.

Kendo

All this has paved the way for fish farmers, from Canada to Vietnam, to go after the most lucrative fish "crop" of all: the big, top predators, such as bluefin tuna, Atlantic cod, and halibut, which command the highest price and are becoming most rare in the wild. The consequences of this for the aquaculture industry—and for our own dinner plates—may be more far-reaching than many of us realize.

# "Reducing" Fish to Produce Fish

For much of the world, particularly the developing world, aquaculture isn't so much about maximizing profit as about having a steady supply of seafood to eat. The vast majority of fish farms worldwide are small-scale, rely on few inputs, and are often closely integrated with crop or livestock production. As a result, these traditional operations make a critical contribution to farmer livelihoods, incomes, and food security.

A recent World Bank survey found that in Vietnam, workers in small-scale catfish-farming operations enjoy higher and more stable incomes, worry less about their daily food source, and are able to send money back to their families.[1] In Bangladesh, rice-fish culture increased incomes by 20 percent and rice yields by 8 percent, while reducing pesticide and fertilizer use.[2] And small-scale tilapia farmers in Central Luzon in the Philippines showed net annual earnings that were 50 percent higher than those of nearby rice farmers.[3]

"The great bulk [of aquaculture] is based on animals feeding low on the food chain," says Sena De Silva with the Network of Aquaculture Centres in Asia-Pacific in Bangkok, Thailand.[4] "It provides an affordable, good-quality animal protein supply to the poor. We feel very perturbed when the Western press talks about aquaculture as totally based on salmon and shrimp."

Yet the reality is that the greatest growth in fish farming today is occurring at the other end of the spectrum: large farms raising high-value, predatory fish. Spurred by lucrative international markets, farming of predatory species such as salmon, striped bass, tuna, and shrimp has expanded by nearly 25 percent a year over the last quarter century.[5] Production of farmed shrimp alone jumped sixfold during this period, while production of farmed salmon increased fourfold.[6] (See Figure 4.)

**Figure 4. Farmed Shrimp and Salmon Production, 1950–2006**

*Source: FAO*

Raising these predatory species is an exercise in "reducing" fish to produce fish—that is, in turning certain fish, usually smaller species such as anchovy, herring, capelin, and whiting, into feed for other, typically larger, species. Increasingly, we are fishing *down* the ocean chain so we can move *up* the fish-farming chain. Fish farmers around the world are also increasingly feeding fish that were traditionally herbivorous with small amounts of fishmeal.[7] Aware of this predicament, the United Nations Food and Agriculture Organization (FAO), in its Code of Conduct for Responsible Fisheries, calls on countries to "encourage the use of fish for human consumption" and discourage its use for feeding animals or other fish.[8]

## "Reducing" Fish to Produce Fish

According to most estimates, modern fish farming is now a net drain on the world's seafood supply.[9] In other words, fish farms that raise species high on the food chain consume considerably more fish in the form of feed than they produce in aquaculture. According to a report from the University of British Columbia's Sea Around Us Project, in 1948, only 7.7 percent of marine fish landings were reduced to fishmeal and fish oil. Today, that share is about 37 percent, eliminating an important historical and future source of human sustenance.[10]

Four groups of fish—marine shrimp, marine fish, trout, and salmon—consume more than half of the world's fishmeal and the vast majority of its fish oil, even though they represent just 7 percent of global aquaculture production and less than 3 percent of total seafood production.[11] Modifying the way these species are produced, or raising less of them, is therefore a priority.

Tuna farming represents one extreme of this evolution. Scientists are just beginning to figure out how to get tuna to spawn and then hatch their eggs in captivity. In the meantime, tuna "ranchers"—in places like Australia, Italy, Mexico, and Spain—rustle up schools of juvenile tuna in the ocean and tow the fish closer to shore using special nets. The tuna are then transfered to grow-out pens anchored in the water and fed with wild-caught sardines until their fat content is high enough to meet market demand.[12] Over the last decade alone, the tuna industry has grown at an astonishing rate, expanding 40 percent in value and 16 percent in volume.[13]

In a seminal 2005 article, food researchers Roz Naylor and Marshall Burke point out that tunas and other "tigers of the sea" have large market potential and "are likely to play a defining role in the future direction of the aquaculture industry."[14] But the long-term viability of tuna ranches seems limited by the fact that these operations depend on dwindling populations of wild fish as a seed stock. Moreover, the conversion of fish feed to harvested fish is very poor: it generally requires 20 kilograms of feed to produce just 1 kilogram of tuna.[15]

There appear to be better ways to do tuna ranching. In contrast to European operations that operate year-round, use imported frozen feed, and create serious coastal pollution problems, tuna ranching in Baja California, off the coast of Mexico, uses fresh, locally caught Pacific sardine as the main feed source. Production is seasonal, and the practice has shown no detectable impact on either the sardine or tuna fisheries.[16] (Still, at the current scale of operations, scientists there have called for a moratorium on any new farms.[17])

Over the years, even fish farmers who raise large predatory species have become more efficient users of feed. Albert Tacon, an aquaculture expert with Aquatic Farms in Hawaii who previously worked for the FAO, notes that as recently as 1997, it took 1.9 kilograms of wild fish to produce 1 kilogram of fed farmed fish on average. By 2001, this ratio had dropped to 1.3 to 1.[18]

There are several reasons for this greater feed efficiency. For one, farmed fish have been bred to grow faster on less feed.[19] In the case of salmon, the production cycle is roughly 20–25 percent shorter today than it was 10 years ago.[20] Fish have also been bred to consume fewer fish, the priciest ingredient in their feed. The fishmeal portion of the salmon diet has dropped from about 60 percent in 1985 to some 35 percent today.[21] Some analysts think a further 50 to 70 percent reduction is possible.[22]

Meanwhile, improved "pelleting" technology allows feed pellets to sink more slowly, enabling nearly all of the food to be ingested rather than wasted. In some cases, this means less pollution in addition to cost savings. On some Norwegian salmon farms, where the feed conversion ratio has been improved dramatically, the amount of excess nitrogen in the water (in the form of wasted feed and fish manure) has decreased from 180 kilograms per ton of fish to approximately 30 kilograms, and solid waste and phosphate releases have also dropped.[23]

And if we must feed fish back to fish, there are better options. For instance, Tacon suggests that fish farmers be more selective in their fishmeal choices and stop using fish that can be eaten directly by humans, such as anchovies or sardines.[24] (See Sidebar 3.) Another possibility is to make greater use of bycatch, or non-target fish that are caught inadvertently while fishers are aiming for other species. Sanctioning the use of bycatch for fish feed, however, may discourage better fishing practices and encourage fishers to pull in more fish that they ostensibly do not want.[25]

Fish farmers could also favor fish scraps or inedible fish species that cannot readily be returned to the ocean. Roughly two-thirds of the Alaskan pollock fishery catch, for example, is destined to become processing byproduct: the remnants are tossed overboard from processing vessels, or the fish oils are burned in diesel generators.[26] But fish "waste" is a relative term. Fish heads, not a big part of the American seafood palate, are often exported to China, where people turn them into stocks, soups, and myriad other nutritious culinary creations.

To avert the looming feed crisis and to take pressure off perfectly edible wild fish, fish farmers could wean themselves from fish-based feed altogether. Several analysts have suggested using slaughterhouse wastes, which represent nearly half the weight of all animals butchered for meat, as a substitute for fish. Although accurate statistics are lacking, an estimated 15–30 million tons of animal byproducts are generated each year, two to three times the amount of fishmeal and fish oil currently produced.[27] (Concerns about mad cow disease have generally encouraged countries to avoid using animal byproducts in animal and fish feed, though so far there have been no cases of disease crossovers.) But even all the world's slaughterhouse waste won't be enough to satisfy the world's rising demand for fish feed if the tremendous projected growth in aquaculture plays out.

Other evidence shows that predatory fish can be raised on a primarily vegetable-based diet, since even the largest fish are more efficient converters of plants and other biomass than pigs and cows.[28] Experimenting with plant-based feeds and moving toward raising more herbivorous species will be in the industry's best interest. As Peter Tyedmers at Dalhousie Uni-

### Sidebar 3. Eating Little Fish to Save Big Ones

It's hard to believe, but eating little fish may be among the innovations that will help safeguard our beleaguered oceans. This is most evident in Peru, where massive schools of small Peruvian anchovy account for about one-tenth of the wild fish netted by fishers worldwide each year. Despite their small size—just a few inches long—each anchovy is chock-full of the same beneficial fatty acids that have made tuna, salmon, and other big fish famous for warding off heart disease and boosting brain development.

And yet, instead of showing up on dinner plates around the world, nearly all of these Peruvian anchovies get ground—or rendered—into fishmeal and fish oil, which will be used to fatten pigs and chickens in factory farms in North America, Europe, Japan, and other meat-producing nations.

At least one Peruvian scientist is questioning the logic of converting precious seafood into low-value feed. Dr. Patricia Majluf, a marine mammal expert and conservationist, and a team of students from the University of Lima, have launched a campaign to change the image of the anchoveta from something that only poor people eat to a fish that could become a tasty dish for well-heeled sophisticates. Majluf convinced the chefs of 30 top Lima restaurants to serve newly created anchoveta dishes—which Peru's president also sampled—all under the glare of local media. In her press appearances, Majluf points to the fact that at a time when malnutrition is still widespread in Peru, people could benefit from eating this nutrient-packed seafood.

Although thousands of Peruvians are employed in the anchovy fishery and processing industry, many more would be able work in a fishery that handled the fish more carefully and packaged it for local human consumption. In fact, the export price of anchovy meal is extremely low. Majluf calculates that if the anchoveta catch were used for human consumption—in anchovy in lemon juice and olive oil, anchovy tempura, anchovy soup, and other traditional recipes—it could generate revenues that are one order of magnitude higher than those presently gained from the export of fishmeal. And Peru could supply both its internal market and a lucrative international market in small fish. The implications of this innovation stretch well beyond South America, since about one-third of the world's fish catch is currently turned into fishmeal.

*Source: See Endnote 24 for this section.*

These anchovies are fresh from the Adriatic Sea.

Jure Šućur

versity in Nova Scotia notes, "aquaculture development policy should explicitly address both the biophysical costs and the limited nature of animal-derived feed ingredients, by encouraging the development of production capacity and markets for low trophic-level species."[29]

Even today, there are cases where fish can be raised almost exclusively on plant-based feeds but are not, simply because farmers or feed producers are not aware of this possibility. For instance, much of the world's farmed sturgeon—which provides a growing share of global caviar as wild sturgeon populations decline—is fed fishmeal, even though the fish is an herbivore. (It's worth noting that the increasing production of farmed sturgeon has not reduced overfishing or poaching of wild sturgeon.[30]) However, recent spikes in the price of grains means that some fish farmers, including catfish growers in the southern United States, can't afford even corn-based feed.[31]

For marine finfish such as salmon, there appears to be no complete substitute for fishmeal and especially for fish oil. Just as feeding grains and slaughterhouse wastes to cows, which are natural grass-eaters, has lead to unforeseen problems such as pathogenic *E. coli* and mad cow disease, attempting to raise salmon on a completely non-fish diet has its own consequences. While predatory fish can survive on plants alone, they don't digest the food as well and generate more waste as a result.[32] And one lifecycle analysis of raising salmon on four different feeds found that while switching from a high-fish diet to a no-fish diet decreased water pollution and took pressure off wild fisheries, it also generated more greenhouse gases since the plant-based ingredients were grown on chemical-intensive farms and trucked from far away.[33]

Despite ongoing improvements, the rapid growth in fish farming in recent decades has effectively outweighed any gains in feeding efficiency. In the case of farmed salmon, the feed conversion ratio has fallen by 25 percent since 1985, but total production has grown by 60 percent.[34] Moreover, farmers have begun to raise species that have much higher feed requirements.[35] As a result, fish farming's share

of the world's fishmeal and fish oil production is soaring. (The other major consumer of these inputs, livestock production, is gradually shifting away from fishmeal and fish oil since farm animals can be raised on a vegetarian diet.)

Aquaculture now consumes 40 percent of the world's fishmeal, up from just 10 percent two decades ago.[36] It is set to outstrip the world fishmeal supply by 2050.[37] Meanwhile, aquaculture feed already consumes over half the world's fish oil and is expected to outstrip supply by 2015.[38] The price of fishmeal has jumped in recent years, and the price ratio of fishmeal to soybean meal has gone from two to three in just the last few years.[39]

Arguably, fish should be using the majority of the world's fish feed and fishmeal, since they convert it much more efficiently than cows, pigs, and chickens.[40] But even fish can't eat feed that doesn't exist. Although aquaculture represents a much more efficient way to raise food than livestock farming, the reality is that the global appetite for farmed fish is putting unsustainable strain on the world's food resources.

Using aquaculture feed more efficiently is important not just because it affects how much seafood the world actually gets to eat, but also because of the resource use implications. Scientists have found that the provision of feed accounts for as much as 90 percent of the total industrial energy inputs used to produce farmed salmon.[41] This includes the energy used to catch and render wild fish into feed, to raise and process chickens (scraps of which are often used in fish feed), to grow and process corn, and to transport all of this to the fish-farming site. (The remaining 10 percent of the energy goes to raising the juvenile fish, transporting the adult fish, and powering the grow-out facilities.[42])

Because of the higher energy costs associated with feeding larger farmed fish such as salmon and shrimp, the efficiency advantage that aquaculture typically has over milk, egg, or even broiler chicken production disappears. In contrast, raising species that are lower on the food chain on plant-based feeds, or raising shellfish on no feed at all, requires two to three times less energy per unit of edible protein produced

**Table 2. Energy Required to Produce Edible Proteins from Aquaculture versus Capture Fisheries and Animal Agriculture**

| Production System | Fossil Fuel Energy Input/ Protein Output |
|---|---|
| **Capture Fisheries** | |
| Shrimp Fisheries | 198 |
| Lobster Fisheries | 192 |
| Pacific Salmon Fisheries | 18–30 |
| King Salmon Fisheries | 40 |
| Atlantic Salmon Fisheries | 29 |
| Cod Fisheries | 20 |
| **Enhanced Fisheries (Ranching)** | |
| Atlantic Salmon Ranching | 7–33 |
| **Agriculture Systems** | |
| Feedlot Beef | 20–78 |
| Swine | 35 |
| Broiler Chickens | 22 |
| Rangeland Beef | 10 |
| Vegetable Row Crops | 2–4 |
| **Aquaculture Systems** | |
| Atlantic Salmon Cages | 50 |
| Catfish | 34 |
| Rainbow Trout Cages | 24 |
| Mussel Longline Culture | 10 |
| Seaweed Culture | 1 |

*Source: See Endnote 43 for this section.*

than most forms of livestock rearing.[43] (See Table 2.) This is because plant-derived inputs, even when grown on a chemical- and energy-intensive farm, are less energy-intensive than fish or animal-derived inputs.[44]

Beyond energy use, feed generates most of the other environmental costs associated with fish farming, including water and air pollution. Lifecycle analyses of nearly every type of production system studied—including Norwegian salmon, Thai shrimp products, and Finnish trout production—support this finding.[45] For freshwater-based rainbow trout production in France, feed production accounted for 52 percent of the total energy use, 82 percent of the contributions to acidification, 83 percent of the greenhouse gas emissions, and 100 percent of "biotic resource use," or the use of natural resources.[46] A general rule of thumb, according to scientists Malcolm Beveridge and David Little, is, "the more external food that is supplied per ton of production, the greater the wastes and the greater the demands on the environment to disperse and assimilate these wastes."[47]

This tradeoff can occur even when raising the same species in different ways.[48] One study found that producing 1 kilogram of tilapia in an intensive, cage-farming system required 1.5 times more industrial energy than raising the same fish in a semi-intensive, pond-farming system, even though the cage system was more productive per unit of area.[49] In fact, while it is generally assumed that farmed fish make more efficient use of resources than wild-caught fish, at least in the case of salmon, one analysis from British Columbia showed salmon farming consumed from 12.7 to 16 hectares of "marine and terrestrial ecosystem support area per metric ton produced," while salmon fishing consumed 5 to 11 hectares.[50] Still, the relative energy efficiency of farmed fish could improve as depleted wild fish schools force boats to motor ever farther from shore.

# Fish Farming for Restoration

In addition to rethinking their use of feed, fish farmers can take further steps to ensure that their operations are environmentally sound and even ecologically restorative. As with agriculture on land, fish farming is often viewed as separate from its natural surroundings. It pollutes waterways, eliminates habitat, and reduces biodiversity. But just as grass-fed cows are generally better for nearby streams than industrial feedlots, and diverse organic vegetable farms typically harbor more songbirds than corn monocultures, some innovative fish farmers are beginning to redesign their farms to function more like healthy aquatic ecosystems.

Consider salmon, the first fish to be farm-raised on a large scale. In 2005, the U.S.-based National Environmental Trust and other conservation groups, fishing organizations, and marine scientists launched an initiative, known as the Pure Salmon Campaign, to address the ecosystem damage associated with large salmon farms.[1] The campaign encourages farms to address eight target issues, including fish waste, the spread of disease, and fish escapes into nearby waters.[2] The campaign is also lobbying the world's largest salmon-farming company, Marine Harvest, which controls 60 percent of global production, with a combination of shareholder resolutions and direct negotiations with corporate boards.[3]

There is reason to be optimistic about large-scale improvements. In Norway, the salmon farming industry has largely eliminated antibiotic use by developing vaccines, rotating sites after each harvest of fish, and imposing strict rules for the movement of live fish.[4] Higher fines for escaped salmon have also pushed Norwegian fish farmers to be more careful about cage and site design and transferring fish. Some salmon farms have eliminated the use of chemical antifoulants by using frequent "swimthroughs" whereby fish are shifted from one place to another—a marine-equivalent of fallowing.

One major thrust of the Pure Salmon Campaign has been pushing for a shift to "closed-container" farms—operations that raise fish in floating containers that are completely separate from the surrounding waters, rather than in the permeable pens that currently predominate. This approach, already in use in some farms around the world, solves many of the problems of waste and fish escapes by essentially isolating the animals within the containers. Such a move admittedly does not confront the high feed requirements of salmon farming, but this larger problem can be softened in part by reconfiguring salmon (and other) farms to include more than just a single species.

Closed-container systems, such as operations that raise fish in concrete or fiberglass tanks using recirculated water, are being tested in a wide range of environments, including large urban areas.[5] (See Sidebar 4.) One problem with these systems, however, is that they tend to require a substantial amount of energy to pump, oxygenate, and cleanse the water, especially when the farms are far from a water source.[6] Unlike raising fish in pens, cages, or open ponds, the tanks aren't able to take advantage of the many services provided by nature, including the tidal action that flushes wastes and replenishes oxygen. And because setting up such farms is expensive, there is pressure to stock fish at great densities, raising

## Sidebar 4. Fish Farms Move from Ocean to Warehouse

At the Center for Marine Biotechnology in Baltimore, Maryland, John Stubblefield and his fellow researchers at the University of Maryland are creating what may be the "next generation" of seafood. Using city-supplied water and a complex microbial filtration system, they are raising a few hundred fish completely indoors. Yonathan Zohar, the center's director and study's leader, says it is the first indoor marine aquaculture system that can re-circulate nearly all of its water and expel zero waste. "I'm a strong believer that in 20 years, most seafood will be grown on land," Zohar says. "It can go to the Midwest, it can go into the inner city, it can go wherever."

If the center's system can become economically competitive with current marine fish-farming techniques, Zohar believes his team may have found a sustainable answer to the world's mounting fisheries crisis. According to some estimates, as much as 90 percent of edible marine fish could be effectively extinct by 2048. The most common alternative to wild fish is fish farms that raise ocean-captured fish in coastal net pens. However, net pens pollute coastal environments with waste and antibiotics, fish escapes pose a threat to the diversity of wild populations, and diseases can spread easily through the fisheries.

Some nations are responding to net-pen pollution by closing troublesome operations. In Israel, for example, the government called for the removal of 2,700 tons of Red Sea net pens by June 2008 due to damage to nearby coral reefs. Zohar spent a decade developing those same net pens when he worked for the Israeli National Center for Mariculture, before relocating to Baltimore in 1990. He says his land-based system is an improved alternative. "[The fish] are disease free, pathogen free; they are contaminant free; they are toxin free," he said. "We tested them. They're as clean as you can get."

Zohar's lab is raising cobia, a high-value fish that can grow to 2 kilograms in eight months. In the wild, the species is found off the eastern coast of North America and in the western Pacific. Cobia do not swim in schools, making them difficult to catch in large amounts, but when raised in an aquaculture operation they become a valuable food product. The lab is growing the cobia faster and more efficiently than in a net pen—about half a kilogram per month, or double the rate of most species, says Stubblefield.

From the start, Zohar's lab was committed to creating a sustainable, low-impact aquaculture system. They say that 99 percent of their water is recycled, with the only losses due to evaporation. An open-air system filled with microbe-covered, honeycomb-shaped plastic detoxifies the ammonia from the water. The water then flows into an oxygen-free system where different bacteria absorb the nitrogen. For the solid fish waste, a separate filter uses microbes to convert the sludge into methane, creating a clean-burning biofuel. The goal is for 10 percent of the aquaculture's energy needs to be offset by the methane byproduct, Zohar says.

While the system offers potential, it still has trade-offs. "When you grow fish in an indoor tank, it takes a fair amount of infra-structure and it can take a fair amount of energy," says Rebecca Goldburg, a senior scientist with the Environmental Defense Fund. Also, the fish being raised are carnivorous, so feeding them requires the input of other fish that are caught or farmed, likely in a less sustainable manner. Several research efforts around the world, including Zohar's lab, are studying whether an algae-based food can replace the food pellets currently used, which are about 40 percent fish meat.

So far, investors have been hesitant to replicate Zohar's aquaculture due to fears that the system cannot compete with net pens. But as seafood demand increases and supply dwindles, Zohar remains confident. "Once the first couple are up and running, this thing is going to spread like fire," he said.

—Ben Block

*Source: See Endnote 5 for this section.*

concerns about disease and fish welfare.

Some farmers have been able to improve on this approach. Jim Carlberg, a marine scientist who runs a 40-year-old aquaculture operation in southern California, produces fish primarily for nearby restaurants, cafeterias, and Asian markets. Using a system of tanks, he raises a domesticated breed of striped bass marketed as California Farmed Striped Bass, selling 90 tons of fresh fish a month. He sells only whole fish, which involves less processing and less waste and allows him to be competitive with cheaper, imported seafood and wild striped bass.

At his location near the Salton Sea, Carlberg takes advantage of geothermal energy, tapping into warm water underground that allows him to keep his tanks at a balmy temperature, without additional heating. This helps the fish grow faster and make better use of their feed, resulting in feed conversion ratios ranging from 1.5 to 2.0. The farm uses algae to improve the water quality, so the wastewater can be reused multiple times before being applied as fertilizer to nearby fields of corn, lettuce, and other high-valued vegetables. "This allows multiple crops to be produced, conserving water, and minimizing fertilizer application and discharge of nutrients to the natural environment," Carlberg explains.[7]

Such systems could be adapted to a wide range of environments and regions. "There is a huge market, a huge potential in urban and suburban areas," says Barry Costa-Pierce,

Barnacle-encrusted mussels offer their ecosystem services off the California coast near Santa Cruz.

Quinn Kuiken

director of the Rhode Island Sea Grant College Program and professor of fisheries and aquaculture at the University of Rhode Island. He knows of farmers with small-scale trout operations in the U.S. state of Vermont who bring all their fish to a nearby farmer's market and "earn more money from their trout every year than from their two acres of vegetables."[8]

In general, aquaculturists have domesticated a much wider range of species than farmers on land—not just plants and mammals, but also mollusks, crustaceans, fish, jellyfish, and worms. According to one recent study, on land, only about 0.08 percent of known plant species and 0.0002 percent of known animal species have been domesticated, compared with 0.17 percent of plants and 0.13 percent of animals in the marine environment—despite the fact that farmers have been selecting species on land for far longer.[9] This wide diversity of marine species lends itself to more integrated

fish farms that use multiple varieties in multiple habitats or at multiple trophic levels.[10]

A more integrated form of fish farming can play an important role in environmental conservation, though this potential remains largely untapped. "To me, the future really has to be integrated aquaculture systems, which use waste from one species to enhance the productivity of another," says Rebecca Goldburg, with the Environmental Defense Fund. "It's not just optimized for the one fish, but for the whole system. Unfortunately, this sort of farming is in its infancy. And that's a problem." [11]

Integrated fish farming can be especially useful in improving water quality. For example, Cooke Aquaculture's operation in Back Bay takes advantage of a natural ecosystem cleansing service provided by shellfish. Because mussels, oysters, clams, scallops, and other shellfish eat algae, a healthy shellfish population can filter and reduce excess nutrients that run into the water. Not only does this help to prevent destructive algal blooms, it has the potential to reverse the expansion of large oxygen-depleted "dead zones" in the world's oceans, more than 200 of which have developed in recent years.[12]

According to estimates, one adult oyster can filter nearly 200 liters of water a day.[13] The billions of oysters that once inhabited places like the Chesapeake Bay in the eastern United States or Puget Sound in the Northwest can help filter an entire bay every few days. This allows sunlight to reach the bay bottom so that grasses and other bases of the food chain thrive. "By providing these three services— filtration, stabilization and habitation—oysters engineered the ecosystem," author Rowan Jacobsen wrote in a 2007 *New York Times* op-ed. He argues that a large-scale return to shellfish farming would yield much more than just jobs and seafood to eat.[14]

Shellfish aren't the only marine species that can play a role in ecosystem cleansing. In China, seaweeds, in addition to mollusks, are often raised in proximity to marine finfish cages to help reduce nutrients and waste from the cages.[15] And one project stocked mullet in small cages under a commercial fish-farm cage to help consume detritus, particular organic matter, unused feed, and bacteria, and to generally reduce the impact on the seafloor. The farming operation will benefit as well. Selectively placed shellfish or seaweed can help prevent fish cages from being colonized, fouled, and damaged by barnacles and algae, which can cause stress in the fish if they impede water flow and compete for food. The practice also reduces the need for costly cage cleaning.

Perhaps of even greater benefit to human populations, integrated fish farming has tremendous potential to clean wastewater, particularly in urban settings. Flows of wastewater in the world's megacities are predicted to increase so greatly in the coming decades that "even with large capital investments in sewage treatment to treat an ever-increasing volume of wastewater, nitrogen loadings to coastal oceans will continue to increase dramatically."[16] And this is assuming that countries can afford these upgrades.

Consider the wetlands outside of Calcutta, India, where 8,000 fish-farm workers manage traditional ponds, called *bheris*, to produce some 13,000 tons of fish a year for the city's 12 million inhabitants. The 3,500 hectares of wetlands are home to many migrating birds. But the bigger environmental service they provide is the fact that the fish feed on the 600 million liters of raw sewage that spews from Calcutta every day, turning a health risk into a valuable urban crop. As a World Bank report notes, "this is the city's sewage treatment plant, deploying a natural cascade of water hyacinth ponds, algal blooms, and fish to dispose of the city's human waste."[17]

In urban Vietnam, the government is promoting raising carp, catfish, tilapia, barb, and Kissing gourami in farm ponds to help clean canals and other surface waters that are contaminated with sewage, including the manure from the many pig and poultry farms cropping up as people eat more meat. The nutrients in the manure stimulate the growth of phytoplankton, zooplankton, and other organisms that ultimately feed the fish. One study found that the highest level of manure inputs led both fish yields and farmer incomes to increase nearly 20-fold compared to the unfertilized system, producing some 8,380 kilograms of fish per year and providing a return of about 52 million Vietnamese dollars (roughly $3,125). The ponds were able to reduce the pollution from pig manure by 60 percent.[18]

In Los Angeles County, California, an aquaculture–wetland ecosystem that grows Chinese water spinach and tilapia for food, and water

Catfish being farmed at a very high density in Vietnam.
Fishwelfare.net

hyacinths for mulch and compost, removed over 97 percent of the nutrients from untreated wastewater, while providing habitat for threatened bird species.[19] It is important to note, however, that these ecosystem services are much easier to harness when industrial and residential wastewater streams are kept separate. In cases where these streams are mixed, the wastewater needs to be extensively cleaned before it can be used to raise fish or plants.

Fish farming can help to restore degraded coral reefs and wetlands as well. The metal cages that hold farmed shellfish often function as artificial reefs around which striped bass, shad, and other marine species congregate.[20] In the Caribbean, the Caicos Conch Farm raises King conch not just to sell to restaurants around the world, but to help re-seed coral reefs with this keystone species.[21]

In the U.S. state of Louisiana, which has lost millions of hectares of wetlands to coastal development and other threats, fish farmers have created 68,000 hectares of red swamp crawfish-rice wetlands, raising the crawfish from autumn to spring and then planting rice in the summer. The habitat created by these farms has encouraged the return of endangered water birds like egrets, herons, ibis, and spoonbills.[22] In the Mississippi delta, a similar recovery has been seen in double-breasted cormorants in proximity to catfish farms.[23] Bird watching, hunting, and ecotourism near fish farms could bring even more revenue to rural areas.

The restorative potential of fish farming is vast and should not be overlooked. The same tools that aquaculturists employ to raise kelp and other plants—using so-called "marine agronomy"—can be harnessed to multiply eelgrass beds, mangrove seedlings, and other lost ecosystems.[24] (This restoration will benefit aquaculture more than most other industries, since it is one of the few industries that requires a constant supply of clean water.) On the Colorado River, where native fish species have been disrupted by damming and other alterations to the water, managers at "ecological hatcheries" attempt to mimic the natural environment of the fish as closely as possible—in terms of feed, genetics, and stocking density—because they realize this offers the best hope that the hatchery fish will survive in the wild.[25]

# A Shift in Mindset

Ultimately, the most basic step in changing our approach to fish farming may be changing our mindset toward it. In a world that depends increasingly on farmed seafood but that faces impending fish feed shortages, this shift will mean favoring species that are lower on the food chain, including seaweeds, shellfish, and herbivorous fish. It will also mean embracing local aquaculture production and supporting scientific research and government policies that promote more sustainable fish farming practices.

In many Asian and European cultures, where people already eat a wide range of seafoods, making the transition to species that are lower on the food chain may not be so traumatic—although eating larger, predatory species remains an important part of the cuisine. In the Americas and in Africa, which have less of a seafood culture, there are fewer traditions to build on. And people seem open to change. Consumption of shellfish, particularly mussels and oysters, has soared in recent years as Americans have been introduced to Asian, European, and other global cuisine.

There are other good reasons to favor species that are lower on the food chain. Like their wild counterparts, certain predatory farmed fish, such as salmon and tuna, carry high risks of contamination that may outweigh the other health benefits they bring.[1] (See Sidebar 5.) Seaweeds in particular have many known health benefits—if the rest of the world were to incorporate seaweed into its diet on par with Asia, there would be an added incentive to integrate sea plants into aquaculture designs.

Also critical to shifting people's attitudes toward fish farming is encouraging greater local acceptance of aquaculture. In parts of Europe that specialize in fish farming—such as Spain, where mussels account for 80 percent of aquaculture production—people understand the important cultural, economic, and nutritional contributions of shellfish farming.[2] It is a locally rooted activity that has spawned food tourism and seafood festivals, including the important Festa do Marisco in Pontevedra each October.[3] Other towns throughout coastal Spain have their own list of seafood festivals, mostly revolving around farmed species, and the practice of aquaculture is widely accepted.

But this isn't true everywhere. Bill Taylor of Taylor Shellfish, whose family has been raising oysters, mussels, geoducks, and other shellfish in the U.S. Pacific Northwest for over 100 years, has watched Americans eat more and more shellfish over the last few decades. Yet while many of the restaurants that buy from him understand the role that shellfish farms play in improving water quality in the Puget Sound, Taylor struggles to find additional shoreline to raise the fish, since new residents don't want farms near their homes. Nonetheless, Taylor has been at the center of efforts to educate homeowners about water quality, create political incentives to upgrade septic systems, and tighten statewide regulations on dumping in the Sound.[4]

Part of the challenge is that some aquaculture industries, such as salmon in Chile and shrimp in Thailand, produce primarily for foreign markets, not local buyers. Whereas farmers of meat and agricultural crops around the world are taking advantage of the growing interest in buying "local food" grown close to

**Sidebar 5. Farmed Fish and Your Health: To Eat or Not to Eat?**

Proponents of aquaculture often argue that because fish farms afford greater control over what the fish eat, the resulting seafood is less likely to be contaminated with heavy metals or other pollutants. While this makes sense theoretically, not all aquaculture is created equal.

For instance, larger carnivorous fish, such as salmon and striped bass, require a diet of small marine fish, which produces the beneficial omega-3 fatty acids in their flesh. On the other hand, herbivorous fish, like tilapia and catfish, are smaller, have short life spans, and aren't fed fishmeal, so they do not accumulate the same amount of fatty acids over time. Since fat-soluble contaminants, such as polychlorinated biphenyls (PCBs), tend to bioaccumulate in the fatty tissue of animals, contaminant levels are highest in the longer-lived, fattier fish species and lowest in herbivorous species.

Contaminant levels vary based on where fish and their feed are caught. The lowest levels of contaminants have been observed in wild and farmed fish from South America; European fish generally have the highest levels, while North American levels fall roughly between the two. Even vegetarian fish raised in polluted water can still accumulate toxins from their environment.

Farmed carnivorous fish are fattier overall in part because their feed is fattier than what their wild counterparts eat. They also do not exert the same energy migrating or catching their food as wild fish. For example, farmed Atlantic salmon contain 1.5 times the fat of wild Atlantic salmon (12 percent versus 8 percent). This fat content also affects the healthfulness of the salmon. While many people eat salmon for its omega-3 fatty acids, farm-raised varieties on average contain less omega-3 (the "good" fatty acid) and more omega-6 (the "bad" fatty acid) than wild salmon, because the fish feed itself contains less omega-3 than the salmon's natural diet. Interestingly enough, however, because farmed salmon is fattier than wild salmon all around, the consumer is still receiving more omega-3 than if her or she ate wild salmon. Yet at the same time, because of their increased fat, the farmed fish also have a greater "body burden" of any environmental contaminants in the feed.

Fish feed, compared to a salmon's natural diet, is more likely to be contaminated with mercury, dioxin-like PCBs, and other persistent organic pollutants. According to one study by the Environmental Working Group, the average farmed salmon has 16 times the PCBs found in wild salmon, 4 times the levels found in beef, and 3.4 times the levels found in other seafood. But while these PCB concentrations may seem high, when compared to the portion size and frequency of consumption of other meats, farmed salmon may not be all that dangerous. In 2006, the U.S. per capita intake of red meat was 50 kilograms, compared to only 7 kilograms for fish and shellfish. Similarly, in 2002, per capita PCB intake from red meat was 2.4 parts per million (ppm), versus only 0.03 ppm for farmed salmon.

In light of these health concerns, the aquaculture industry is taking greater control of the fish diet. Farmers are shifting from feeds high in contaminants to feeds with lower levels. To meet this demand, one of the largest fishmeal manufacturers in Europe announced in 2005 that it was building a facility to extract dioxin from fishmeal. Large North American producers are taking similar steps.

Another popular alternative is moving toward plant-based feeds, which tend to be lower in fat and are less likely to be contaminated. But such a shift isn't without consequence, since plants also have much lower levels of the omega-3 fatty acids that have become the signature of healthy seafood. In a recent study in Norway, heart-disease patients fed different varieties of farmed salmon showed the best results from eating salmon that contained very high levels of omega-3 fatty acids (generally found in fishmeal and fish oil), and benefited considerably less from salmon fed on a diet of rapeseed oil containing much lower omega-3 levels. To solve this problem, aquaculturalists have started feeding salmon a solely vegetarian diet up until the last few months before the harvest, when fish are fed purely a marine feed. This practice can restore omega-3 levels in salmon to 80 percent of their natural content.

—Hannah Doherty

*Source: See Endnote 1 for this section.*

home, aquaculture's bounty often doesn't show up in neighborhood fish markets. Involving the local community in these operations could go a long way toward improving fish farming's image.

Several years ago, the residents of Charlotte County, New Brunswick, now one of the cen-

ters of Canadian salmon farming, realized that for every job created on a fish farm, another 3–5 jobs are involved in producing feed and processing and marketing the fish. The community decided to invest in local feed mills, local shops to make and service equipment, and a local processing facility. Nearly one-quarter of the residents are now employed in the industry, and, in a good sign for the future, 75 percent of employees are below the age of 40. Even as the total number of jobs has increased, the share of jobs that are full-time has risen from 60 to 80 percent.[5]

In Egypt, after the government identified aquaculture as the best hope for closing the gap between seafood supply and demand, production soared from 35,000 tons in 1997 to 471,535 tons today.[6] The country's 70 million people eat an average of 14.7 kilograms of seafood each year, roughly twice as much as other nations in Africa.[7] (On the continent as a whole, where seafood consumption is actually declining, aquaculture has tremendous potential as a source of local food and livelihood; see Sidebar 6.[8])

About half of Egypt's production is tilapia, which thrives in the abundant brackish delta lakes and lagoons. After investors and university researchers took notice of aquaculture's potential, feed mills grew from just two in 1997 to more than 32, and hatcheries jumped from 14 in 1998 to 520. Interestingly, even as the industry has grown, small farms of 2–4 hectares have begun to outnumber the more traditional large farms of 50–200 hectares, because they yield a better return per unit of land. And very little of the national production is exported, being consumed almost entirely by locals.[9]

People who eat from their local waters have a natural reason to be concerned about what goes into them. "Spat" is the term for baby shellfish, but it also stands for the Southold Program in Aquaculture Training (SPAT), an initiative launched in 2000 by Cornell University's Marine Center on New York's Long Island to provide additional firepower for local shellfish restoration.[10] For a $150 start-up fee, community volunteers receive spat and the equipment necessary to raise the creatures in floating cages. They also get an entire year of graduate-level training in the nuances of algae growth, marine ecology, and shellfish dynamics—as well as a chance to restore the popular scallop economy. Participants report that they end up changing their daily habits that affect

## Sidebar 6. Africa's Untapped Fish Farms

Globally, fish farming's biggest untapped potential is in Africa, the only region in the world where seafood consumption is actually dropping. Projections show that just to maintain Africa's per capita food-fish consumption of roughly 8 kilograms a year, sub-Saharan Africa's harvest must increase from roughly 6.2 millions tons a year today to 9.3 million tons in 2020. Assuming that wild fish catches remain the same (a precarious assumption considering how much of the African catch is expropriated by legal and illegal fleets from Europe and elsewhere), fish farming would have to jump nearly fourfold over the next couple of decades.

This sounds like a lot, but it's about the same rate of growth as global aquaculture production over the last couple of decades. As a World Bank report notes, "there are no physical and technological barriers to a major expansion of sustainable aquaculture in sub-Saharan Africa." An estimated 30 percent of the region's land has the needed water for small-scale fish farming. Currently, just 7 percent of sub-Saharan Africa's fish comes from farms, compared with more than 40 percent worldwide and roughly 70 percent in Asia.

Asia, in contrast, has benefited from decades of research, extension, and scientific support. Fish Farmers' Development Agencies in 22 Indian states trained more than 550,000 farmers and introduced a range of technologies that helped boost carp polyculture in more than 450,000 hectares of ponds; as a result, production of Indian carps increased from 50 kilograms per hectare to about 2,200 kilograms between 1974 and 1999. Municipalities in China and Vietnam modified zoning laws to remove barriers to adding fish to rice farming, while in Bangladesh, members of landless communities were granted access to public canals and other water bodies to raise fish.

Many of these innovations can be transplanted to Africa, but the continent needs its own infrastructure to support it. The World Bank points to "evidence of a sea change" in recent years. In Malawi, under a program established by the U.S. Agency for International Development, farmers now set aside a small amount of their land for fish farming in ponds fed by farm and kitchen wastes, including maize bran, household leftovers, and animal manure. Compared to traditional crop farms, these integrated systems show a sixfold increase in profits, and crop yields are 18 percent higher.

The ponds help farmers cope with drought and enable them to raise crops like cabbage and tomatoes that normally require irrigation during the dry season. In families that have added aquaculture to their farming operation, child malnutrition has dropped by some 15 percent. "Fish in the pond is like money in the bank," says Jessie Kaunde, a farmer and widow in Mangwengwe village in southern Malawi. Some 5,000 farmers have adopted the system in Malawi and Zambia.

*Source: See Endnote 8 for this section.*

water quality, such as shunning chemical fertilizers from their lawns, upgrading their home septic systems, and using nontoxic paints on their boats.[11]

Unlike similar programs started in the

Chesapeake Bay and elsewhere, SPAT provides an attractive gastronomic incentive: gardeners get to eat half their harvest of fresh, mature shellfish. The other half goes to Cornell University's marine center and hatchery in Southold, which in turn distributes the mollusks regionwide to bring back the glory of the Peconic Bay, once a center of scallop, oyster, and clam production. There are signs that the program is working: baymen, naturalists,

In this photo from the International Space Station, sunglint off the water highlights aquaculture empoundments along the northeast coast of the Nile delta in Egypt.

Courtesy of Earth Sciences and Image Analysis Laboratory, NASA Johnson Space Center

and hobby anglers have all noticed significantly more baby scallops than in recent memory. Dozens of shoreline communities, from Cape Cod to Chile, hope to replicate the SPAT program.

A similar shift in mindset needs to take place in the seafood industry and among aquaculture specialists. As we use more of our streams, lakes, and bays for farming, it is important to learn from the mistakes that land-based agriculture has made. On land, the farming industry, agricultural researchers, and decision makers are only beginning to view agriculture as a venture that must be integrated with conservation goals, reduce its dependence on chemical and external inputs, and embrace diversity. Consider all the rapidly growing number of agricultural schools around the world that have carved out a department of sustainable agriculture. "The science of aquaculture is only about 30 years old, so we have a

long way to go," says Sena De Silva, director general of the Network of Aquaculture Centres in Asia-Pacific in Bangkok.

One important step is to ensure that aquaculture operations are developed in a manner that is locally appropriate. While much of the world's fish farming is based on exotic species—salmon in Chile, for instance—De Silva and others feel this is an accident of history. Some researchers are beginning to look at indigenous species that can be cultured, including Indian and Chinese carps and catfishes, in an effort to reduce a farm's chemical use and susceptibility to disease.[12] (Another benefit of such research is that, on all continents, aquaculture has often been the culprit in introducing invasive species that wreak havoc on local fish, mollusks, plants, and aquatic food webs.)

"It's amazing that, even in Asia, aquaculture is not taught as a true interdisciplinary science," notes Barry Costa-Pierce with the Rhode Island Sea Grant College Program. "There is no mass movement of government or industry to see that happen…. We don't have the next generation of ecological designers and integrated thinkers in this field."[13] As in agriculture, an obstacle to more integrated fish farming is the trend toward greater specialization, with more farms focused on just one species or one phase of production (whether hatchery or pre-growing or growing out), without considering all the stages of the life cycle.[14]

Costa-Pierce and Canadian researcher Thierry Chopin are now involved in forming a global network of Sustainable Ecological Aquaculture Systems (SEAS) labs that have commercial-scale examples of ecological, multitrophic aquaculture and that are ready to share this knowledge across continents. The group has begun to speak with large fish-farming concerns, as well as with land-based agriculture operations that could benefit from greater integration with fish farms (whether to produce feed for livestock or fertilizer for fields), with the main idea that pollutants are merely "misplaced resources." Such thinking is still essential, says Costa-Pierce: "We have a chance with aquaculture that we never had with agriculture to make it a totally sustainable integrated industry worldwide."[15]

Governments can play at least some role in making this shift happen. In the same way countries are beginning to tax certain agriculture operations for the pollution or other costs they place on society, they might guide fish farmers by penalizing farms that pollute and rewarding farms that clean up. Such an approach may meet resistance at first, but it would gradually become accepted. Several decades ago, when tuna fishers were forced by public pressure to reduce the number of dolphins killed during fishing, the industry first balked. But before long, fishers were able to modify their practices, and the changes were quickly adopted by much of the world's tuna fleet.

Another option could be to grant "nutrient credits," similar to carbon credits that a nation or business might receive for reducing its emissions of climate-altering greenhouse gases, to farms that incorporate nitrogen-sequestering shellfish or kelp into their design. Governments could also provide incentives to encourage other ocean-based industries, from offshore wind farms to tidal-energy plants, to use aquaculture in their design as a way to make better use of existing infrastructure and to reduce their own ecological footprint.

Unfortunately, this sort of government guidance often comes slowly, and often long after the broader public has realized the need to seek out a different food in the marketplace. This points to a greater role for certification and standards to spur the industry to develop more sustainable practices and products.

# Toward Sustainable Aquaculture Standards

It's hard to believe that the aquaculture industry is already providing nearly half of our seafood, yet there are no widely accepted standards for what constitutes "good" fish farming. This includes fish farming that limits pollution, improves nearby habitats, and minimizes food safety risks. For comparison, the organic food industry has strong international and national standards, even though it constitutes just 3 to 5 percent of the world's food supply.

Recent legislation in both Europe and the United States requires mandatory certification to identify whether seafood is wild or farm-raised. Europe has standards for organic aquaculture, and the United States is working on similar standards. But there is still no widespread labeling or other consumer information on how farmed fish are produced.

In 2003, the United Nations Food and Agriculture Organization formed a new working group on an "Ecosystems Approach to Aquaculture" to build on its earlier Code of Conduct for Responsible Fisheries.[1] "That's a huge policy leap," says Barry Costa-Pierce, who with Thierry Chopin is one of the group's leaders. "Governments worldwide have to learn to say no to unsustainable aquaculture. Mozambique is going to say, 'here's this giant shrimp farm coming from outside the country.' They have to have the capacity to say no if it cannot be done in a socially and environmentally responsible manner."[2]

Fortunately, some developing countries with large export markets have responded to the demand for standards from industrial-nation buyers. In Bangladesh, certain farmed shrimp and prawns can carry the Shrimp Seal of Quality label, while the Thai Department of Fisheries has developed a Thai Quality Shrimp label. Malaysia has standards for all farmed species, and in Chile, the world's second largest salmon producer behind Norway (and the nation with the most potential for growth), Fundación Chile and SIGES-Salmon Chile have developed the Code of Good Environmental Practices to certify the fish.[3] Even voluntary standards can spur significant change in the industry, since farms that don't sign up can end up learning from the standards.[4] Belize, Colombia, and Madagascar have all considered using certification to differentiate their products in the global marketplace.

And the industry itself is getting the message. "Aquaculturists are increasingly aware that it is in their own best interests to adopt sustainable practices to reduce problems with pollution and disease," says Daniel Lee, Best Aquaculture Practices Coordinator with the Global Aquaculture Alliance (GAA), a group founded in 1997 that now has 1,100 members in 70 countries. As of October 2006, GAA's accreditation body had certified 50 processing plants, 26 farms, and 17 hatcheries on three continents, including in most major exporting nations.[5] In contrast to the wild and more-difficult-to-govern fishing industry, Lee believes, "farm owners have an incentive to invest in the long-term viability of their operations, knowing that the benefits of good management will not be dissipated by outsiders."[6]

Currently, GAA just certifies shrimp and shrimp feed producers, but standards for tilapia, salmon, pangasius, and catfish are in the works. The Alliance chose to focus on shrimp in response to the soaring international trade in shrimp in the 1980s and early 1990s, which led producers in South and East Asia, who supply the majority of the world's shrimp, to greatly expand shrimp areas. Hundreds of thousands of hectares of mangrove forests and coastal rice paddies were cleared or converted for shrimp farms. The effluent from intensive shrimp ponds often contains high amounts of nutrients, excess feed, chemicals, and drugs. In the last decade, governments and shrimp growers have gradually begun to replant many of the mangroves and to reintegrate rice growing into shrimp production, but only after realizing that wholesale landscape conversion was polluting the water, exacerbating disease, and generally reducing shrimp production.[7]

Under the GAA's approach, getting certified depends on meeting a variety of standards to achieve a minimum score, although certain critical actions, including "those dealing with mangroves, effluents, antibiotics and hatchery seed," are mandatory. In addition, the GAA has a set of guiding principles that encourage—but don't require—farms to continually reduce their feed and medicine use, improve their water quality, and share the economic benefits with local communities.[8]

In this sense, the GAA standard is an immensely practical one. Consider that some organic fish farming standards, such as those drafted by Naturland of Germany and the Soil Association of the United Kingdom, place restrictions on stocking densities—that is, how many fish can be kept in a given area. The GAA takes the view that "such restrictions are arbitrary" and that more intensive farms might actually make more efficient use of water, feed, and seed.[9] The standards make regular use of the word "realistic," and the speed with which they were developed and implemented has allowed adoption by even the largest seafood buyers, including Darden Restaurants, the parent company of Red Lobster, which with 1,300 locations is the top seafood restaurant chain in the United States. In early 2006, Darden announced plans to certify all of its farm-raised shrimp.[10]

Meanwhile, some environmental groups have criticized the GAA's standards as being too weak, alleging that they stop short of significant environmental safeguards to instead allow producers a lower hurdle for gaining compliance. As one alternative, the conservation group WWF is attempting to establish internationally recognized standards for some 11 important farmed fish and shellfish, an ambitious project that involves input from fish farmers, marine scientists, consumer advocates, and environmental groups. WWF hopes the standards will attract the interest of large food buyers like Whole Foods Market and Carrefour.[11]

Farmed shrimp from Halong Bay, Vietnam.

Jeff Butterworth

WWF is building on its success in developing standards for forestry (the Forest Stewardship Council) and wild fisheries (the Marine Stewardship Council)—both of which are now widely used and have moved their respective industries away from polluting and destroying biodiversity and toward using more recycled materials and enhancing ecosystem health.[12] The goal of the new aquaculture standards is to "minimize or eliminate the environmental and social impacts responsible for 70 to 80 percent of the problems caused by aquaculture."[13] The effort has grown in part out of WWF's own conclusion that shrimp farming can be improved more easily than shrimp fishing, a process that often involves ocean-scouring nets that scoop up more unsuspecting fish, turtles, and other ocean life than they do shrimp.

The standards, due out by the end of 2009, are aimed at Europe, the United States, and Japan, the three leading shrimp buyers.[14] But they also target major shrimp producers, such as Brazil, Thailand, and Vietnam, as well as China and India, which are both producing and eating more shrimp. The "science-based" standards will ultimately include measurable goals for a farm's performance, such as reduced impact on mangroves, less reliance on wild shrimp larvae for seed, and finding substitutes for fishmeal. But there is also an interest in making them practical and ensuring that they are adopted widely by industry and others. While the fish farming industry is involved in developing the standards, the ultimate decision lies with a diverse group, unlike with the GAA, where industry makes the ultimate decision.[15]

More recently, the Pew Environment Group launched an effort to come up with its own certification stands, in part out of a concern that both the industry and WWF efforts were flawed (the industry approach because the standard-setters didn't seem sufficiently removed from the companies and farmers, and the WWF approach because it was constrained by the need for consensus among its diverse stakeholders). Pew is less concerned about whether its standards can be used by the industry: "We are after precautionary science-based standards, a gold standard," explains Chris Mann, senior officer and director of the Campaign for Healthy Oceans at the Pew Environment Group. "The idea of science-based standards is not whether they are achievable but where we want to be."[16]

It's interesting to note that the Marine Stewardship Council (MSC), the certification body that administers the global ecolabel for wild seafood, has decided to stay out of the farmed fish debate. (It could play a role indirectly, however, if it certifies the wild anchovy and sardine and other baitfish fisheries that feed aquaculture.) But as some of its bigger seafood-buying clients seek MSC's device, chief executive Rupert Howes says the group is carefully considering whether to throw its hat into the ring, knowing that its reputation could

help establish credibility for certified farmed seafood and maintain clarity in labeling.[17]

Still, MSC's 10 years of experience have demonstrated the power to change the market. In the last two years, MSC has seen an almost fourfold increase in the number of fisheries in the program and products in the marketplace, including 40 percent of the global whitefish catch and 40 percent of all wild salmon. "As we get more products in [the] marketplace, brand recognition increases and so does understanding of the label," says Howe. "Now less well-managed fisheries are coming to the program and they realize they need to make significant changes, and that's when we can make the real impact."[18]

Large food companies whose brands depend on sustainability have had to seek out their own farmed fish and develop their own standards, often by partnering with a conservation group. In the United Kingdom, both the Marks & Spencer and Sainsbury's chains have started selling only salmon that meets the "Freedom Fish" standards developed by the Royal Society for the Protection of Cruelty to Animals; the standards include giving the fish 20 percent more space than typical farms, which means they grow more slowly and have less fat, and lowering their body temperature in a tank to make them less aware of their slaughter.[19]

On the other side of the Atlantic, the U.S. food chain Whole Foods Market recently announced the first comprehensive set of aquaculture guidelines by a major retailer, prompted mostly by "customers' expectations," according to the company.[20] The standards, developed over two years with input from scientists and conservation groups, apply to all frozen, fresh, canned, and smoked seafood (except mollusks) and include prohibitions on preservatives, antibiotics, hormones, and other chemicals that are typically used to control disease and encourage growth in fish. They also ban Whole Foods from buying from fish farms in wetlands and mangroves and limit how much wild fish can be used to feed farmed fish.

In the case of Wegmans supermarkets in the northeastern United States, a Wegmans food safety expert worked with fisheries expert Rebecca Goldburg at the Environmental Defense Fund to develop the company's policies for farmed salmon and shrimp.[21] Perhaps most importantly, any producer in the Wegmans program must keep data on its use of feed and veterinary medicines, and its water usage and discharges. The suppliers must also track whether ponds are allowed to dry out between shrimp cycles to help keep down disease and pursue several other metrics that allow the farm's impact on the surrounding area to be monitored and ultimately improved.

"A lot of it is information that they already have," explains Teresa Ish, seafood project manager for Wegmans' corporate partnerships program.[22] "The fact that the report is publicly available is also really important. If all farms reported this type of information, it would make such a big difference in conservation because it would set farms up to compete with each other to be the best farm." Such growing practices also mean a superior taste and texture, according to feedback from customers. The chain's shrimp, produced primarily in Central America, is billed as a "green" product that is still affordable—more expensive than the frozen shrimp from Asia, but less expensive than fresh wild shrimp from U.S. boats.

Bon Appétit, the U.S. food service company with 400 cafés at colleges, universities, and corporate campuses in 28 states, also worked with the Environmental Defense Fund to develop shrimp standards. The company already had among the most advanced seafood buying policies in the industry, tied to the Monterey Bay Aquarium's seafood watch program. (It buys nothing on the program's "red list" and actively works to promote "green-list" items to its customers and suppliers.) For its East Coast cafes, the company is looking into a New Brunswick salmon farmer, since its low-carbon diet program deemed the footprint of its wild frozen Pacific salmon too great. By 2009, the company will eliminate all air-freighted seafood, and it is actively introducing low trophic-level species like mussels, oysters, sardines, and tilapia to its menus.

In the case of shrimp, used in everything from cold salads to stir-frys, the complex supply chain meant that it was almost impossible for Bon Appétit's supplier to provide the sort of information it wanted. So the company worked for nearly a year with a Texas shrimp producer, whose product was smaller than Asian shrimp and considerably more expensive because of the smaller scale of the farm. Despite the fact that Americans love shrimp, Bon Appétit has decided to treat the shellfish as a special, limited-supply item—"like local, heirloom tomatoes."[23] Where chefs can find

An aerial view of an inland shrimp farm in Belize.
Cathy Chaput

a local source, they are encouraged to do so. But Bon Appétit has reduced the presence of shrimp on its menus by as much as 50 percent companywide.

Certification standards are just one of the battlegrounds as we try to steer half of the world's seafood in another direction, says Astrid Scholz, an ecological economist at the conservation organization Ecotrust. Scholz has been heading up an international project to dissect the economic and environmental impact of farmed fish.[24] "We're taking a cradle to grave approach," she explains, from the source of the feed and how it gets to the farm, to the electricity used to filter farm water. "We're not just considering proximal ecological impact, sea lice, or pollution, but also what happens in processing, transportation, and what is the carbon footprint."[25]

For the hypothetical consumer in New York, Paris, or Beijing that is browsing the frozen seafood section, the important question now is: what is the correct choice, especially if frozen fillets from wild and farmed fish are indistinguishable? "You can't just blithely say, 'everybody go ahead and eat wild salmon,' when it means for a European customer that salmon arrives by air from North America. We would really like people to understand that there's more to seafood sustainability than what goes on in the water," says Scholz.[26]

Scholz's research aims to ultimately guide certification schemes or regulatory efforts, which are focusing primarily on the fish itself and on what goes on in the water. The good news, she says, is that "it becomes very apparent that systems can be considerably improved."

In this sense, there is a great convergence between fish farming and rising concerns about meat, vegetables, and other things we eat. Just as buyers of meat and milk are becoming more vigilant, seafood shoppers will also begin to ask more questions about what the fish are fed, how they are raised, and where they are from—either because they wonder if it makes the fish less healthy to eat, or because they are concerned it harms the fish.[27]

"What is the simple consumer message?" Scholz asks. "There are no simple answers when it comes to seafood anymore."

# Endnotes

## A Different Sort of Salmon

1. Thierry Chopin, University of New Brunswick, St. John, New Brunswick, Canada, discussion with author, 23 March 2008; Shawn Robinson, University of New Brunswick, St. John, New Brunswick, Canada, discussion with author, 17 April 2008. See also Thierry Chopin, "Integrated Multi-trophic Aquaculture (IMTA) Will Also Have Its Place When Aquaculture Moves to the Open Ocean," *Fish Farmer*, March/April 2008, and Susan Llewelyn Leach, "Growing Seaweed Sustainably," *Gulf of Maine Times*, Winter/Spring 2008.

2. Chopin, discussion with author, op. cit. note 1.

3. Ibid.

4. United Nations Food and Agriculture Organization (FAO), FAOSTAT electronic database, faostat.fao.org, updated May 2008.

5. FAO, *The State of World Fisheries and Aquaculture 2006* (Rome: 2007).

6. World Bank, *Changing the Face of the Waters: The Promise and Challenge of Sustainable Aquaculture* (Washington, DC: 2007), pp. 1–2.

7. Marie Christine Monfort, *Markets and Marketing of Aquaculture Finfish in Europe: Focus on the Mediterranean Basin*, report prepared for FAO Fisheries division (Rome: April 2006), pp. 22–23; Josh Eagle, Rosamond Naylor, and Whitney Smith, "Why Farm Salmon Outcompete Fishery Salmon," *Marine Policy*, May 2004, pp. 259–70.

8. Figure 1 from FAO, FISHSTAT electronic database, www.fao.org/fishery/topic/16073, viewed March 2008.

9. Ibid.

10. FAO, op. cit. note 8.

11. FAO, op. cit. note 4.

12. John Volpe, School of Environmental Studies, University of Victoria, British Columbia, Canada, discussion with author, 3 April 2008.

13. Projection from FAO, op. cit. note 5; other statistics from FAO, op. cit. note 8.

14. U.S. Department of Commerce, "Development of a Code of Conduct for Responsible Aquaculture in the United States Exclusive Economic Zone. Public Workshops, 24 August 2000," *Federal Register*, Vol. 65, No. 165 (2000), pp. 51591–92; U.S. Department of Commerce, "U.S. Department of Commerce Aquaculture Policy," www.nmfs.noaa.gov/trade/DOCAQpolicy.htm, 8 April 2001.

15. Carlos M. Duarte, Nùria Marbá, and Marianne Holmer, "Rapid Domestication of Marine Species," *Science*, 20 April 2007, pp. 382–83.

16. Melba G. Bondad-Reantaso et al., "Disease and Health Management in Asian Aquaculture," *Veterinary Parasitology*, 30 September 2005, pp. 249–72.

17. Figure 2 from FAO, op. cit. note 8.

18. Chopin, discussion with author, op. cit. note 1.

19. Ibid.

20. Ibid.

21. Ibid.

22. Janice Harvey and Inka Milewski, *Salmon Aquaculture in the Bay of Fundy: An Unsustainable Industry* (Fredericton, New Brunswick, Canada: Conservation Council of New Brunswick, September 2007), at www.conservationcouncil.ca/files/PDF/Aquaculture/Salmon_Aquaculture.pdf.

23. Robinson, op. cit. note 1.

24. Chopin, discussion with author, op. cit. note 1.

25. Ibid.

26. Ibid.

27. World Bank, op. cit. note 6.

## From Ornamental Ponds to Industrial Aquaculture

1. Malcolm C.M. Beveridge and David C. Little, "The History of Aquaculture in Traditional Societies," in Barry Costa-Pierce, ed., *Ecological Aquaculture: The Evolution of the Blue Revolution* (Oxford, UK: Blackwell, 2003), p. 7.

2. Ibid.

3. Ibid.

4. Ibid., p. 12.

5. Ibid., p. 11.

6. Ibid., p. 13.

7. Ibid., p. 13.

8. Ibid., p. 14.

**9.** Ibid., p. 30.

**10.** Ibid., p. 8.

**11.** Ibid., p. 35.

**12.** WorldFish Center, "GIFT Fish," www.worldfishcenter .org/cms/list_article.aspx?catID=32&ddlID=76, updated 28 November 2997.

**13.** WorldFish Center, "Rice-Fish Culture: A Recipe for Higher Production," www.worldfishcenter.org/cms/list _article.aspx?catID=32&ddlID=78, updated 28 November 2007.

**14.** World Bank, *Changing the Face of the Waters: The Promise and Challenge of Sustainable Aquaculture* (Washington, DC: 2007).

**15.** Beveridge and Little, op. cit. note 1, p. 17.

**16.** Ibid., p. 11.

**17.** Figure 3 from United Nations Food and Agriculture Organization (FAO), FISHSTAT electronic database, www.fao.org/fishery/topic/16073, updated March 2008.

**18.** Ibid.

**19.** Ibid.; Albert G.J. Tacon, "State of Information on Salmon Aquaculture Feed and the Environment," report prepared for WWF-US, unpublished report shared with author.

**20.** FAO, op. cit. note 17.

**21.** Ibid.

**22.** Rebecca Goldburg, Environmental Defense Fund, New York, NY, discussion with author, 14 March 2008; FAO, op. cit. note 17.

**23.** Beveridge and Little, op. cit. note 1, p. 37.

**24.** Sidebar 1 from the following sources: M. Beveridge, *Cage Aquaculture* (Oxford, UK: Blackwell Publishers, Fishing News Books, 1996); M. MacGarvin, "Scotland's Secret? Aquaculture, Nutrient Pollution, Eutrophication, and Toxic Blooms," *Modus Vivendi* (Perth, Scotland: WWF Scotland, 2000); Rosamond Naylor and Marshall Burke, "Aquaculture and Ocean Resources: Raising Tigers of the Sea," *Annu. Rev. Environ. Resour.*, Vol. 30 (2005), pp. 185–218; spending on disease and losses from Melba G. Bondad-Reantaso et al., "Disease and Health Management in Asian Aquaculture," *Veterinary Parasitology*, 30 September 2005, pp. 249–72; Martin Krkošek et al., "Epizootics of Wild Fish Induced by Farm Fish," *Proceedings of the National Academies of Sciences*, 17 October 2006, pp. 15506–10; John Volpe, School of Environmental Studies, University of Victoria, British Columbia, Canada, discussion with author, 3 April 2008; tuna disease from John P. Volpe, "Dollars Without Sense: The Bait for Big-Money Tuna Ranching Around the World," *BioScience*, April 2005, pp. 301–02.

**25.** Beveridge and Little, op. cit. note 1, p. 34.

**26.** FAO, op. cit. note 17; Peter Tyedmers et al., "Biophysical Sustainability and Approaches to Marine Aquaculture Development Policy in the United States," a report to the Marine Aquaculture Task Force (Takoma Park, MD: Marine Aquaculture Task Force, February 2007).

**27.** Henning Roed, WWF-Norway, discussion with author, 26 March 2008; SeaWeb Aquaculture Resources, "SeaWeb History of Salmon Farming," www.seaweb.org/ resources/aquaculturecenter, viewed 10 April 2008; Naylor and Burke, op. cit. note 24; Dietrich Sahrhage and Johannes Lundbeck, *A History of Fishing* (Berlin: Springer-Verlag, 1992).

**28.** Naylor and Burke, op. cit. note 24.

**29.** Ibid.

**30.** Roed, op. cit. note 27.

**31.** Ibid.

**32.** Ibid.

**33.** World Bank, op. cit. note 14.

**34.** Table 1 from ibid.

**35.** John P. Volpe, "'Salmon Sovereignty' and the Dilemma of Intensive Atlantic Salmon Aquaculture Development in British Columbia," in C.C. Parrish, N.J. Turner, and S.M. Solberg, eds., *Resetting the Kitchen Table: Food Security, Culture, Health and Resilience in Coastal Communities* (Hauppauge, NY: Nova Science Publishers, Inc., 2006).

**36.** Naylor and Burke, op. cit. note 24, p. 190.

**37.** Ibid.

**38.** Tacon, op. cit. note 19.

**39.** Naylor and Burke, op. cit. note 24, pp. 209–10.

**40.** Volpe, op. cit. note 35.

**41.** Ibid.; Volpe, discussion with author, op. cit. note 24.

**42.** Ussif Rashid Sumaila, John Volpe, and Yajie Liu, "Potential Economic Benefits from Sablefish Farming in British Columbia," *Marine Policy*, Vol. 31 (2007), pp. 81–84.

**43.** Volpe, "Dollars Without Sense…," op. cit. note 24.

**44.** Volpe, op. cit. note 35.

**45.** Beveridge and Little, op. cit. note 1, p. 25.

**46.** Naylor and Burke, op. cit. note 24.

**47.** Sidebar 2 from the following sources: Bondad-Reantaso et al., op. cit. note 24; Roed, op. cit. note 27; Naylor and Burke, op. cit. note 24; "Is Farmed Fish Not All Bad? Kona Blue Raises Its Fish With Sustainability in Mind," *The Daily Green*, www.thedailygreen.com, 3 April 2008; Neil Sims, President/Co-founder, Kona Blue, Kailua-Kona, HI, discussion with author, 26 April 2008; Nick Joy, Managing Director, Loch Duart Ltd., "*Sustainability and the Future of Seafood*," presentation at Seafood Summit 2006, Seattle, WA, 29–31 January 2006; Canadian Science Advisory Secretariat, "Assessing Potential Technologies for Closed Containment Saltwater Salmon Aquaculture," Science Advisory Report 2008/001 (Ottawa, ON: Fisheries and Oceans Canada, March 2008); Chris Mann, Senior Officer and Director, Campaign for Healthy Oceans, Pew Environment Group, Washington, DC, discussion with author, 19 December 2007.

# Endnotes

## "Reducing" Fish to Produce Fish

1. World Bank, *Changing the Face of the Waters: The Promise and Challenge of Sustainable Aquaculture* (Washington, DC: 2007).

2. Ibid.

3. Ibid.

4. Sena De Silva, Network of Aquaculture Centres in Asia-Pacific (NACA), Bangkok, Thailand, discussion with author, 4 February 2008.

5. United Nations Food and Agriculture Organization (FAO), FISHSTAT electronic database, www.fao.org/fishery/topic/16073, updated March 2008.

6. Figure 4 from ibid.

7. Ibid.

8. FAO, "Code of Conduct for Responsible Fisheries," www.fao.org/docrep/005/v9878e/v9878e00.HTM.

9. Rosamond Naylor and Marshall Burke, "Aquaculture and Ocean Resources: Raising Tigers of the Sea," *Annu. Rev. Environ. Resour.*, Vol. 30 (2005), pp. 185–218.

10. Jackie Alder and Daniel Pauly, "On the Multiple Uses of Forage Fish: From Ecosystems to Markets," *Fisheries Centre Research Reports* (Sea Around Us Project, University of British Columbia), Vol 14, No. 3 (2006), pp. vii, 3.

11. Albert G.J. Tacon, "State of Information on Salmon Aquaculture Feed and the Environment," report prepared for WWF-US, unpublished report shared with author.

12. John P. Volpe, "Dollars Without Sense: The Bait for Big-Money Tuna Ranching Around the World," *BioScience*, April 2005, pp. 301–02.

13. Naylor and Burke, op. cit. note 9, p. 193.

14. Ibid.

15. Ibid., p. 195; Volpe, op. cit. note 12.

16. Jose A. Zertuche-Gonzalez et al., "Marine Science Assessment of Capture-based Tuna (*Thunnus orientalis*) Aquaculture in the Ensenda Region of Northern Baja California, Mexico," Final Report of the Binational Scientific Team to the Packard Foundation (Stamford, CT: Department of Ecology & Evolutionary Biology, University of Connecticut, 2008).

17. Ibid.

18. Tacon, op. cit. note 11.

19. Ibid.

20. Ibid.

21. Ibid.

22. Ibid.; Albert Tacon, Aquaculture Research Director, Aquatic Farms Ltd., Kaneohe, HI, discussion with author, 8 January 2008.

23. Tacon, op. cit. note 11.

24. Ibid. Sidebar 3 from the following sources: Patricia Majluf, University of Lima, Peru, discussion with author, 2 February 2008; Daniel Pauly, "Babette's Feast in Lima," *The Sea Around Us Project Newsletter*, November/December 2006; Danielle Nierenberg and Brian Halweil, "Farming the Cities," in Worldwatch Institute, *State of the World 2007* (New York: W.W. Norton & Company, 2008.)

25. Tacon, op. cit. note 11.

26. Neil Sims, President/Co-founder, Kona Blue, Kailua-Kona, HI, discussion with author, 26 April 2008.

27. Tacon, op. cit. note 11.

28. Ibid.

29. Peter Tyedmers et al., "Biophysical Sustainability and Approaches to Marine Aquaculture Development Policy in the United States," a report to the Marine Aquaculture Task Force (Takoma Park, MD: Marine Aquaculture Task Force, February 2007), p. 27.

30. Phaedra Doukakis, Pew Institute for Ocean Science, New York, NY, discussion with author, 12 November 2007; "Increase in Caviar Quota Sparks Anger," fishupdate.com, 2 June 2008.

31. David Streitfeld, "As Price of Corn Rises, Catfish Farms Dry Up," *New York Times*, 18 July 2008.

32. Tacon, op. cit. note 11.

33. N. Pelletier and P. Tyedmers, "Feeding Farmed Salmon: Is Organic Better?" *Aquaculture*, Vol. 272 (2007), pp. 399–416.

34. Naylor and Burke, op. cit. note 9.

35. Ibid.

36. Tacon, op. cit. note 11.

37. Ibid.

38. Ibid.

39. Ibid.

40. World Bank, op. cit. note 1, p. 32.

41. Tyedmers et al., op. cit. note 29; C. Folke, "Energy Economy of Salmon Aquaculture in the Baltic Sea," *Environmental Management*, Vol. 12, No. 4 (1988), pp. 525–37; M. Troell et al., "Aquaculture and Energy Use," in C. Cleveland, ed., *Encyclopedia of Energy, Vol. 1* (San Diego: Elsevier, 2004), pp. 97–108.

42. Tyedmers et al., op. cit. note 29.

43. Table 2 from Barry A. Costa-Pierce, "Ecology as the Paradigm for the Future of Aquaculture, in Barry Costa-Pierce, ed., *Ecological Aquaculture: The Evolution of the Blue Revolution* (Oxford, UK: Blackwell, 2003), pp. 339–72.

44. Tyedmers et al., op. cit. note 29.

45. Ibid.

46. Ibid.

47. Malcolm C.M. Beveridge and David C. Little, "The History of Aquaculture in Traditional Societies," in Costa-Pierce, ed., op. cit. note 43, p. 5.

48. Pelletier and Tyedmers, op. cit. note 33.

49. Tyedmers et al., op. cit. note 29.

50. Ibid.

## Fish Farming for Restoration

**1.** Don Staniford, Pure Salmon Campaign, Oslo, Norway, discussion with author, 21 November 2007. See also www.puresalmon.org.

**2.** Ibid.

**3.** Ibid.

**4.** Melba G. Bondad-Reantaso et al., "Disease and Health Management in Asian Aquaculture," *Veterinary Parasitology*, 30 September 2005, pp. 249–72; Henning Roed, WWF-Norway, discussion with author, 26 March 2008.

**5.** A recent overview of current commercial use of closed-container systems is Ecoplan International Inc., *Global Assessment of Closed System Aquaculture*, prepared for David Suzuki and Georgia Strait Alliance (Vancouver, British Columbia: May 2008). Sidebar 4 adapted from Ben Block, "New Fish Farms Move from Ocean to Warehouse," *Eye on Earth* (Worldwatch Institute), 25 April 2008; 90 percent from Boris Worm et al., "Impacts of Biodiversity Loss on Ocean Ecosystem Services," *Science*, 3 November 2006, pp. 787–90; net pens from United Nations Environment Programme, *GEO Yearbook 2006* (Nairobi: 2006), p. 67; 10 percent from United Nations Food and Agriculture Organization (FAO), *The State of World Fisheries and Aquaculture 2006* (Rome: 2007); Israel from "Eliat's Fish Farms: In? Out?" *Israel Environment Bulletin*, January 2005; methane from U.S. Environmental Protection Agency, "Methane," www.epa.gov/methane.

**6.** Rosamond Naylor and Marshall Burke, "Aquaculture and Ocean Resources: Raising Tigers of the Sea," *Annu. Rev. Environ. Resour.*, Vol. 30 (2005), pp. 185–218.

**7.** James M. Carlberg, President, Kent SeaTech Corp, San Diego, CA, discussion with author, 5 May 2008.

**8.** Barry Costa-Pierce, Director, Rhode Island Sea Grant College Program, University of Rhode Island, Kingston, RI, discussion with author, 3 and 10 April 2008 and 14 June 2008.

**9.** Carlos M. Duarte, Nùria Marbá, and Marianne Holmer, "Rapid Domestication of Marine Species," *Science*, 20 April 2007, pp. 382–83.

**10.** Ibid.

**11.** Rebecca Goldburg, Environmental Defense Fund, New York, NY, discussion with author, 14 March 2008.

**12.** J. Michael Beman, Kevin R. Arrigo, and Pamela A. Matson, "Agricultural Runoff Fuels Large Phytoplankton Blooms in Vulnerable Areas of the Ocean," *Nature*, 10 March 2005, pp. 211–14.

**13.** Rowan Jacobsen, *A Geography of Oysters* (New York: Bloomsbury USA, 2007).

**14.** Rowan Jacobsen, "Restoration on the Half Shell," *New York Times*, 9 April 2007.

**15.** World Bank, *Changing the Face of the Waters: The Promise and Challenge of Sustainable Aquaculture* (Washington, DC: 2007).

**16.** Barry A. Costa-Pierce and Christopher J. Bridger, "The Role of Marine Aquaculture Facilities as Habitats and Ecosystems," in R.R. Stickney, ed., *Responsible Marine Aquaculture* (Oxfordshire, UK: CABI International, 2002).

**17.** World Bank, op. cit. note 15.

**18.** Dang K. Nhan et al., "Economic and Nutrient Discharge Tradeoffs of Excreta-fed Aquaculture in the Mekong Delta, Vietnam," *Agriculture, Ecosystems and Environment*, Vol. 124 (2008), pp. 259–69.

**19.** Costa-Pierce, op. cit. note 8; Kenneth R. Weiss, "A Primeval Tide of Toxins," *Los Angeles Times*, 30 July 2006.

**20.** Jacobsen, op. cit. note 14.

**21.** Chuck Hesse, Caicos Conch Farm, Turks and Caicos, discussion with author, 28 January 2008. See also www.caicosconchfarm.com.

**22.** Costa-Pierce and Bridger, op. cit. note 16.

**23.** Ibid.

**24.** Ibid.

**25.** Costa-Pierce, op. cit. note 8.

## A Shift in Mindset

**1.** Rosamond Naylor and Marshall Burke, "Aquaculture and Ocean Resources: Raising Tigers of the Sea," *Annu. Rev. Environ. Resour.*, Vol. 30 (2005), pp. 185–218. Sidebar 5 from the following sources: antibiotic use from Charles Benbrook, "Antibiotic Drug Use in U.S. Aquaculture," February 2002, at www.mindfully.org/Water/Antibiotic -Aquaculture-BenbrookFeb02.htm, and from Chuck Benbrook, Northwest Science and Environmental Policy Center, Sandpoint, ID, e-mail to author, 16 June 2008; Nick Jans, "Farmed Salmon Can't Beat Wild," *USA Today*, 6 October 2002; contaminants in farmed salmon from Environmental Working Group, *PCBs in Farmed Salmon: Factory Methods, Unnatural Results* (Washington, DC: 30 July 2003); dioxin extraction facilities from Triple Nine Fish Protein company, *Triple Nine News*, No. 2 (2005), available at www.999.dk; health study from S. L. Seierstad et al., "Dietary Intake of Differently Fed Salmon: The Influence on Markers of Human Atherosclerosis," *European Journal of Clinical Investigation*, 5 January 2005, pp. 52–59; contaminant levels by region from R.A. Hites et al., "Global Assessment of Organic Contaminants in Farmed Salmon," *Science*, Vol. 303 (2004), pp. 226–29; fat in Atlantic salmon from Department of Foods and Nutrition, Purdue University, "Fatty acid content of farmed and wild fish," at http://fn.cfs.purdue.edu/ang lingindiana/AquaculturevsWildFish/FattyAcidsFarm.pdf; red meat vs. fish intake from U.S. Department of Agriculture, Economic Research Service, "Red meat, poultry, and fish (boneless, trimmed equivalent)," electronic database, www.ers.usda.gov/Data/FoodConsumption/ FoodAvailSpreadsheets.htm, updated 15 March 2008; 2003 salmon consumption from H.M. Johnson and Associates, *Annual Report on the United States Seafood Industry* (Jacksonville, OR: 2004); 2002 PCB intake from Ron Hardy, "Contaminants in Salmon: A Follow-Up," *Aquaculture Magazine*, March/April 2005, pp. 1–3; 80 percent of omega-3 content from G. Bell et al., "Dioxin and Dioxin-like Polychlorinated Biphenyls (PCBs) in Scottish Farmed Salmon (*Salmo salar*): Effects of Replacement of Dietary Marine Fish Oil with Vegetable Oils," *Aqua-

# Endnotes

*culture*, Vol. 243 (2005), pp. 305–14.

**2.** United Nations Food and Agriculture Organization (FAO), FISHSTAT electronic database, www.fao.org/fishery/topic/16073, updated March 2008.

**3.** Galicia Tourism Guide online, "Fiesta del Marisco," www.galinor.es/ogrove/.

**4.** Bill Taylor, Taylor Shellfish Farms, Shelton, WA, discussion with author, 23 March 2008.

**5.** Government of New Brunswick, Canada, Department of Agriculture and Aquaculture, *Aquaculture 2005*, at www.gnb.ca/0168/30/ReviewAquaculture2005.pdf; Len Stewart, Aquaculture Strategies Inc., "Salmon Aquaculture in New Brunswick: Natural Development of Our Marine Heritage," report prepared for The New Brunswick Salmon Growers' Association (New Brunswick, Canada: December 2001).

**6.** Egypt from Marie Christine Monfort, *Markets and Marketing of Aquaculture Finfish in Europe: Focus on the Mediterranean Basin*, report prepared for FAO Fisheries division (Rome: April 2006), p. 32.

**7.** Ibid.

**8.** Ibid. Sidebar 6 from the following sources: World Bank, *Changing the Face of the Waters: The Promise and Challenge of Sustainable Aquaculture* (Washington, DC: 2007); Christopher Delgado et al., *Outlook for Fish to 2020: Meeting Global Demand* (Washington, DC: International Food Policy Research Institute, 2003); World Fish Center, "Waste Not Want Not: Spreading Affordable Integrated Agriculture-Aquaculture in Malawi, Sub-Saharan Africa," www.worldfishcenter.org/cms/list_article.aspx?catID=328&ddlID=77, viewed 4 September 2006; FAO, op. cit. note 2; Randall Brummet, WorldFish Center, Penang, Malaysia, discussion with author, 17 September 2007.

**9.** Monfort, op. cit. note 6.

**10.** Brian Halweil, "Salty Saviors," *Orion*, July/August 2005.

**11.** Ibid.

**12.** Sena De Silva, Network of Aquaculture Centres in Asia-Pacific (NACA), Bangkok, Thailand, discussion with author, 4 February 2008.

**13.** Barry Costa-Pierce, Director, Rhode Island Sea Grant College Program, University of Rhode Island, Kingston, RI, discussion with author, 3 and 10 April 2008 and 14 June 2008.

**14.** Monfort, op. cit. note 6.

**15.** Costa-Pierce, op. cit. note 13.

## Toward Sustainable Aquaculture Standards

**1.** United Nations Food and Agriculture Organization (FAO), "Promoting the Ecosystem Approach to Fisheries. Guidelines on the Ecosystem Approach in Fisheries to Promote Sustainable and Responsible Fisheries," www.fao.org/bestpractices/content/06/06_03_en.htm; Barry Costa-Pierce, "An Ecosystem Approach to Marine Aquaculture: A Global Review," in D. Soto et al., eds., *Building an Ecosystem Approach to Aquaculture (EAA): Initial Steps for Guidelines: FAO Working Group on an Ecosystem Approach to Aquaculture Expert Workshop, 7–11 May, 2007. Mallorca, Spain* (Rome: FAO, 2008).

**2.** Barry Costa-Pierce, Director, Rhode Island Sea Grant Colege Program, University of Rhode Island, Kingston, RI, discussion with author, 3 and 10 April 2008 and 14 June 2008.

**3.** Daniel Lee, Global Aquaculture Alliance, St. Louis, MO, discussion with author, 28 March 2008.

**4.** Ibid.

**5.** Lee, op. cit. note 3.

**6.** Ibid.

**7.** Mai Trong Thong et al., "An Assessment of the Linkages Between Trade Liberalization, Rural Poverty, and Environment in Shrimp Aquaculture in Ca Mau Province, Vietnam" (Hanoi: WWF Greater Mekong Vietnam Programme, October 2006); Institute of Fisheries Management, Research Institute for Aquaculture Number 1, Network of Aquaculture Centres in Asia-Pacific, Can Tho University, and World Wide Fund for Nature, "Guidelines for Environmental Management of Aquaculture Investments in Vietnam," Technical Note 37564, prepared for the Ministry of Fisheries Viet Nam and for the World Bank (Washington, DC: June 2006).

**8.** Lee, op. cit. note 3.

**9.** Ibid.

**10.** "Buyers Navigate Sustainable Seafood," *SeaFood Business*, October 2004, pp. 22–23; Kate Moser, "Sustainable Seafood Casts a Wider Net," *Christian Science Monitor*, 3 April 2006.

**11.** Kris Hudson and Wilawan Watcharasakwet, "The New Wal-Mart Effect: Cleaner Thai Shrimp Farms," *Wall Street Journal*, 24 July 2007

**12.** WWF, "Aquaculture Dialogues," www.worldwildlife.org/what/globalmarkets/aquaculture/item5218.html.

**13.** Ibid.

**14.** Eric Bernard, shrimp aquaculture dialogue leader, WWF, Paris, discussion with author, 3 May 2008.

**15.** Ibid.

**16.** Chris Mann, Senior Officer and Director, Campaign for Healthy Oceans, Pew Environment Group, Washington, DC, discussion with author, 19 December 2007.

**17.** Rupert Howes, Chief Executive, Marine Stewardship Council, London, UK, discussion with author, 17 February 2008.

**18.** Ibid.

**19.** Martin Hickman, "The Salmon Business: Can Marine Farming Ever Be Eco Friendly?" *The Independent*, 24 July 2008.

**20.** Ylan Q. Mui, "Grocers' Rules Follow Wave of Sustainably Farmed Fish," *Washington Post*, 16 July 2008. Standards available at Whole Foods Market, "Our Farm-Raised Seafood," www.wholefoodsmarket.com/products/seafood/aquaculture.html.

# Endnotes

**21.** Rebecca Goldburg, Environmental Defense Fund, New York, NY, discussion with author, 14 March 2008.

**22.** Teresa Ish, Environmental Defense Fund, New York, NY, discussion with author, 14 March 2008.

**23.** Helene York, Director, Bon Appétit Management Company Foundation, San Francisco, CA, discussion with author, 20 March 2008.

**24.** Astrid Scholz, Vice President of Knowledge Systems, Ecotrust, Portland, OR, discussion with author, 17 April 2008. See also www.ecotrust.org/lca.

**25.** Ibid.

**26.** Ibid.

**27.** Ibid.

# Index

# Index

# Index

pesticides, 12
Pew Environmental Group, 15, 34
Philippines, 12, 17
pollock, 19
predatory fish, 16, 17–21, 19, 28. *See also specific species*
preservatives, 35
Puget Sound, 24
Pure Salmon Campaign, 22

## R

Red Lobster, 33
Rhode Island Sea Grant College Program, 24
rice-fish culture, 12, 17, 29
Robinson, Shawn, 7
Roed, Henning, 14
Royal Society for the Protection of Cruelty to
    Animals, 34

## S

sablefish, 13
Sainsbury's, 34
salmon
    certification, 22, 32, 34
    contamination, 27, 28
    energy inputs, 20, 21
    farming volume, 17
    fishmeal consumption, 18
    history of farming, 14–15
    integrated multitrophic aquaculture, 7, 10
    veterinary costs, 9
    waste production, 13
Salton Sea (California), 23
sardines, 19, 34
scallops, 29
Scholz, Astrid, 36
Scotland, 13, 16
sea lice, 9, 13
seafood
    certification, 9, 10, 32–36
    consumption rate, 7, 8, 29
    farmed *vs.* wild, 14, 36
    production, 8, 12, 14
    supply, 18, 29
    *vs.* fish feed, 17–21
seaweed
    2006 production, 9
    as waste filter, 7
    domestication, 8
    energy inputs, 21
    health benefits, 27
    integrated multitrophic aquaculture, 24, 25
sewage treatment, 25
shellfish
    as waste filters, 7, 24, 25
    consumption, 27
    energy inputs, 20
    restoration, 29–30
    standards, 31

shrimp
    certification, 32–34, 35–36
    disease losses, 13
    environmental costs, 21
    fishmeal consumption, 18
    production figures, 9, 14, 17
Shrimp Seal of Quality, 32
SIGES-Salmon Chile, 32
Sims, Neil, 15
Skretting, 15
slaughterhouse waste, 19
Soil Association, 33
South Korea, 12
Southold Program in Aquaculture Training (SPAT), 29
Spain, 27
standards, 32–36
Stubblefield, John, 23
sturgeon, 20
Sustainable Ecological Aquaculture Systems (SEAS), 30
swine, 21

## T

Tacon, Albert, 18
Taylor, Bill, 27
Thai Quality Shrimp, 32
Thailand, 12, 27, 32, 33
tilapia
    2006 production, 9
    certification, 32
    early Egyptian aquaculture, 11
    economic benefits, 17
    energy inputs, 21
    fatty acid accumulation, 28, 29
    price and production, 14
    wastewater treatment, 25
trout, 18, 21
tuna
    "ranching," 18
    contamination, 27
    dolphin mortality, 31
    health benefits of, 19
    herpes virus, 13
Tyedmers, Peter, 19

## U

UN Food and Agriculture Organization (FAO), 17, 32
US Agency for International Development (USAID),
    29
US Department of Commerce, 8

## V

vaccines, 22
vegetables, 21
Vietnam, 12, 25, 29
Volpe, John, 15

## W

wakame, 9

Worldwatch Reports provide in-depth, quantitative, and qualitative analysis of the major issues affecting prospects for a sustainable society. The Reports are written by members of the Worldwatch Institute research staff or outside specialists and are reviewed by experts unaffiliated with Worldwatch. They are used as concise and authoritative references by governments, non-governmental organizations, and educational institutions worldwide.

### On Climate Change, Energy, and Materials

175: Powering China's Development: the Role of Renewable Energy, 2007
169: Mainstreaming Renewable Energy in the 21st Century, 2004
160: Reading the Weathervane: Climate Policy From Rio to Johannesburg, 2002
157: Hydrogen Futures: Toward a Sustainable Energy System, 2001
151: Micropower: The Next Electrical Era, 2000
149: Paper Cuts: Recovering the Paper Landscape, 1999
144: Mind Over Matter: Recasting the Role of Materials in Our Lives, 1998
138: Rising Sun, Gathering Winds: Policies To Stabilize the Climate and Strengthen Economies, 1997

### On Ecological and Human Health

174: Oceans in Peril: Protecting Marine Biodiversity, 2007
165: Winged Messengers: The Decline of Birds, 2003
153: Why Poison Ourselves: A Precautionary Approach to Synthetic Chemicals, 2000
148: Nature's Cornucopia: Our Stakes in Plant Diversity, 1999
145: Safeguarding the Health of Oceans, 1999
142: Rocking the Boat: Conserving Fisheries and Protecting Jobs, 1998
141: Losing Strands in the Web of Life: Vertebrate Declines and the Conservation of Biological Diversity, 1998
140: Taking a Stand: Cultivating a New Relationship With the World's Forests, 1998

### On Economics, Institutions, and Security

173: Beyond Disasters: Creating Opportunities for Peace, 2007
168: Venture Capitalism for a Tropical Forest: Cocoa in the Mata Atlântica, 2003
167: Sustainable Development for the Second World: Ukraine and the Nations in Transition, 2003
166: Purchasing Power: Harnessing Institutional Procurement for People and the Planet, 2003
164: Invoking the Spirit: Religion and Spirituality in the Quest for a Sustainable World, 2002
162: The Anatomy of Resource Wars, 2002
159: Traveling Light: New Paths for International Tourism, 2001
158: Unnatural Disasters, 2001

### On Food, Water, Population, and Urbanization

176: Farming Fish for the Future, 2008
172: Catch of the Day: Choosing Seafood for Healthier Oceans, 2007
171: Happer Meals: Rethinking the Global Meat Industry, 2005
170: Liquid Assets: The Critical Need to Safeguard Freshwater Ecosytems, 2005
163: Home Grown: The Case for Local Food in a Global Market, 2002
161: Correcting Gender Myopia: Gender Equity, Women's Welfare, and the Environment, 2002
156: City Limits: Putting the Brakes on Sprawl, 2001
154: Deep Trouble: The Hidden Threat of Groundwater Pollution, 2000
150: Underfed and Overfed: The Global Epidemic of Malnutrition, 2000
147: Reinventing Cities for People and the Planet, 1999

**To see our complete list of Reports, visit www.worldwatch.org/taxonomy/term/40**

The Worldwatch Institute is an independent research organization that works for an environmentally sustainable and socially just society, in which the needs of all people are met without threatening the health of the natural environment or the well-being of future generations. By providing compelling, accessible, and fact-based analysis of critical global issues, Worldwatch informs people around the world about the complex interactions among people, nature, and economies. Worldwatch focuses on the underlying causes of and practical solutions to the world's problems, in order to inspire people to demand new policies, investment patterns, and lifestyle choices.

Financial support for the Institute is provided by the American Clean Skies Foundation, the Blue Moon Fund, the Casten Family Fund of the Chicago Community Trust, the Compton Foundation, Inc., The Goldman Environmental Prize, the Jake Family Fund, the W. K. Kellogg Foundation, the Steven C. Leuthold Family Foundation, the Marianists of the USA Sharing Fund, the V. Kann Rasmussen Foundation, The Shared Earth Foundation, The Shenandoah Foundation, the Sierra Club, Stonyfield Farms, the TAUPO Fund, the Flora L. Thornton Foundation, the United Nations Environment Programme, the United Nations Population Fund, the Wallace Genetic Foundation, Inc., the Wallace Global Fund, the Johanette Wallerstein Institute, and the Winslow Foundation. The Institute also receives financial support from many individual donors who share our commitment to a more sustainable society.